Managing Performance and Assets; Freight Data and Visualization 2018

Transportation Research Record: Journal of the Transportation Research Board Volume 2672, Issue 44

ISBN: 978-1-5108-8250-8

Printed from e-media with permission by:

Curran Associates, Inc.
57 Morehouse Lane
Red Hook, NY 12571

Some format issues inherent in the e-media version may also appear in this print version.

Published by Sage Publications USA

ISBN (Print) 978-1-5108-8250-8
ISBN 2018 Printed Set (All Issues) 978-1-5108-7335-1

Printed by Curran Associates, Inc. (2019)

Additional copies of this publication are available from:

Curran Associates, Inc.
57 Morehouse Lane
Red Hook, NY 12571 USA
Phone: 845-758-0400
Fax: 845-758-2633
Email: curran@proceedings.com
Web: www.proceedings.com

Volume 2672 Number 44 December 2018

TRR
JOURNAL OF THE TRANSPORTATION RESEARCH BOARD

Contents

Managing Performance and Assets; Freight Data and Visualization

Articles

---

* Membership as of February 2019.

TRB

TRANSPORTATION RESEARCH BOARD

CORE SPONSORS

Transportation Departments of the 50 States and the District of Columbia

Federal Government

U.S. Department of Transportation
- Federal Aviation Administration (FAA)
- Federal Highway Administration
- Federal Motor Carrier Safety Administration (FMCSA)
- Federal Railroad Administration (FRA)
- Federal Transit Administration
- National Highway Traffic Safety Administration (NHTSA)
- Office of the Assistant Secretary of Transportation for Research and Technology (OST-R)

U.S. Air Force Civil Engineer Center (AFCEC)
U.S. Coast Guard (USCG)
U.S. Department of Energy
U.S. Department of the Interior (DOI)
U.S. Environmental Protection Agency (EPA)

Nongovernmental Organizations

American Public Transportation Association (APTA)
Association of American Railroads (AAR)
California Air Resources Board

The National Academies of

SCIENCES · ENGINEERING · MEDICINE

The **National Academy of Sciences** was established in 1863 by an Act of Congress, signed by President Lincoln, as a private, nongovernmental institution to advise the nation on issues related to science and technology. Members are elected by their peers for outstanding contributions to research. Dr. Marcia McNutt is president.

The **National Academy of Engineering** was established in 1964 under the charter of the National Academy of Sciences to bring the practices of engineering to advising the nation. Members are elected by their peers for extraordinary contributions to engineering. Dr. C. D. Mote, Jr., is president.

The **National Academy of Medicine** (formerly the Institute of Medicine) was established in 1970 under the charter of the National Academy of Sciences to advise the nation on medical and health issues. Members are elected by their peers for distinguished contributions to medicine and health. Dr. Victor J. Dzau is president.

The three Academies work together as the National Academies of Sciences, Engineering, and Medicine to provide independent, objective analysis and advice to the nation and conduct other activities to solve complex problems and inform public policy decisions. The Academies also encourage education and research, recognize outstanding contributions to knowledge, and increase public understanding in matters of science, engineering, and medicine.

Learn more about the National Academies of Sciences, Engineering, and Medicine at **www.national-academies.org.**

The **Transportation Research Board** is one of seven major programs of the National Academies of Sciences, Engineering, and Medicine. The mission of the Transportation Research Board is to increase the benefits that transportation contributes to society by providing leadership in transportation innovation and progress through research and information exchange, conducted within a setting that is objective, interdisciplinary, and multimodal. The Board's varied activities annually engage about 7,000 engineers, scientists, and other transportation researchers and practitioners from the public and private sectors and academia, all of whom contribute their expertise in the public interest. The program is supported by state transportation departments, federal agencies including the component administrations of the U.S. Department of Transportation, and other organizations and individuals interested in the development of transportation.

Learn more about the Transportation Research Board at **www.TRB.org**.

Article

TRR
JOURNAL OF THE TRANSPORTATION RESEARCH BOARD

Transportation Research Record
2018, Vol. 2672(44) 1–9

DOI: 10.1177/0361198118757981
journals.sagepub.com/home/trr
SAGE

A Semi-Automated Method to Generate GIS-Based Sidewalk Networks for Asset Management and Pedestrian Accessibility Assessment

Hanyan Li[1], Jack Cebe[1], Sara Khoeini[2], Yanzhi "Ann" Xu[1], Chelsea Dyess[1], and Randall Guensler[1]

Abstract

Pedestrian infrastructure that is comfortable, connected to destinations of interest, and accessible to those with disabilities is vital to a safe, accessible, equitable, and sustainable transportation system. Planners recognize the benefits of providing well-maintained sidewalks and curb ramps, but often lack the asset management systems necessary to inventory sidewalk maintenance problems, prioritize sidewalk maintenance needs, and track the implementation of sidewalk improvement projects. Communities that are managing sidewalk presence and condition data typically link the data to their roadway network, which makes tracking specific sidewalk assets difficult. This paper introduces an affordable, semi-automated, and easy-to-implement process to generate a GIS-based sidewalk network with associated links and nodes representing crosswalks and intersections. Quantitative sidewalk condition data can be loaded onto the network, which allows it to be used to manage sidewalks as transportation assets, assessing pedestrian accessibility, prioritizing repairs or improvements, and to automatically identify accessible routes between origins and destinations. System inputs include parcel-level land-use and roadway centerline data, both of which are publicly available and free in most cases. The network is generated within the ArcGIS environment, using Python scripts to implement embedded ArcGIS functions. The method requires few computational resources, and tremendously reduces the manual labor required to develop a fully interconnected sidewalk network. Examples from multiple communities are presented to show how quantitative sidewalk condition data are loaded onto the network, and illustrate the network's potential for pedestrian navigation and routing applications.

Sidewalk infrastructure is a critical component of sustainable transportation systems. Almost every trip made on the transportation network begins and ends with a walking trip on a sidewalk. Good quality pedestrian infrastructure also supports walking as a viable primary mode choice and extends the effective service area of transit networks. Presence and quality of sidewalks are significant predictors of perceived safety and general satisfaction in the pedestrian environment (*1*). Mode shifts to walking (and other non-motorized forms of transportation) have also been associated with increases in physical activity, improved health (*2*), and potential decreases in vehicle miles of travel, roadway congestion, energy consumption, and emissions (*3*). Sidewalks and curb ramps must meet design standards under the Americans with Disabilities Act (ADA) of 1990. Furthermore, case law recognizes that sidewalks are a vital component of accessible transportation infrastructure designed to improve the quality of life of persons with disabilities (*4*). Although local agencies must develop ADA Transition Plans, and may be legally responsible for inadequate infrastructure maintenance, it is often a challenge for jurisdictions to collect the data necessary to measure and assess compliance due to a lack of resources.

Jurisdictions that currently track sidewalk assets use a variety of methods of defining sidewalk features in geographical information sciences (GIS), such as tying sidewalk data to an established roadway centerline network or manually tracing sidewalk links using aerial photography. These methods have drawbacks, namely geo-spatial accuracy, the time required to generate features, and network analysis capability. For example, in a roadway centerline database where sidewalks exist on both sides of the street, two sidewalk links must be attached to each roadway link (as well as indexing the sidewalk to indicate on which side of the roadway it is

[1]School of Civil and Environmental Engineering, Georgia Institute of Technology, Atlanta, GA
[2]School of Sustainable Engineering and the Built Environment, Arizona State University, Tempe, AZ

Corresponding Author:
Address correspondence to Hanyan Li: hli378@gatech.edu

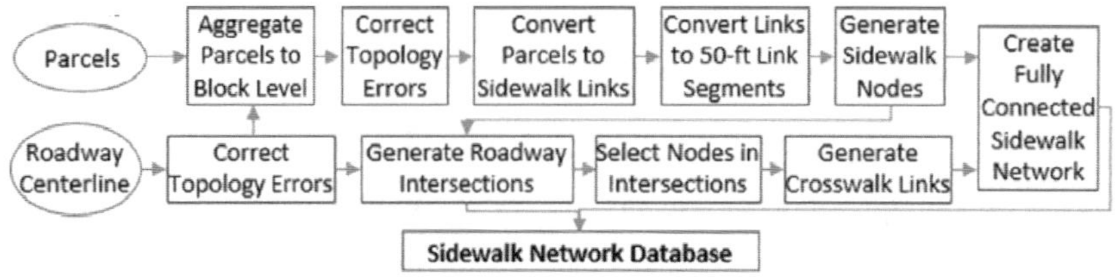

Figure 1. Flow chart of generating sidewalk network.

located). Furthermore, one node typically represents each intersection in the roadway centerline database. However, a four-way intersection can have four or more crosswalks and eight or more curb ramps. It is difficult to manage all of the condition data for ramps and other elements at a single roadway intersection node.

This paper presents an affordable, semi-automated, GIS-based methodology to generate an independent sidewalk network database, with links and nodes readily available to carry condition data for pedestrian infrastructure features, such as sidewalks, crosswalks, curb cuts, and ramps. The network structure and database can be used to assess pedestrian accessibility and prioritize pedestrian infrastructure maintenance projects. The paper is organized as follows: Section 2 introduces previous studies about generating sidewalk networks. Section 3 describes the details of the semi-automated methodology. Section 4 evaluates the method with regard to labor costs. Section 5 uses examples from Northampton, Massachusetts, and East Point, Georgia that show how sidewalk design and condition data are integrated into the sidewalk network. The result is a sidewalk asset inventory and a sidewalk network useful for pedestrian routing. The last section discusses conclusions and regulatory recommendations.

Literature Review

Several studies have generated independent sidewalk networks, mainly within the field of GIS. Karimi and Kasemsuppakorn presented an extensive review of sidewalk network map generation approaches (*5*). The three primary approaches to sidewalk network generation cited in the literature included network buffering, collaborative mapping with global positioning system (GPS) traces, and image processing. The outcome of all three approaches depends on the completeness and quality of the input data. Ballester et al. presented a method based on images to generate sidewalks and walkable plains (i.e., squares) (*6*). Kim et al. made a distinction between roads and crossings and point out that network accuracy hinges on the sophistication of the spatial datasets (*7*). Parker and Vanderslice proposed a tool to create sidewalks from road centerlines and create crossings based on road intersections (*8*). A recent study suggested that sidewalk network distances would provide more realistic pedestrian distances for evaluating walkability and accessibility than distances based upon roadway centerline data (*9*). In previous sidewalk generation research reported by Georgia Institute of Technology (Georgia Tech), the team used Perl script to develop sidewalk networks, but the accuracy was low when generating crosswalk links (*10*). Of the previous approaches that have been employed, network buffering required the least computation effort, and was the least time-consuming and resource-intensive approach (image processing was the most computationally intensive and expensive). However, networks developed using buffering methods tend to be of lower quality and require data that are often not readily available to jurisdictions and agencies.

Methodology

The semi-automated method presented in this paper includes two elements: 1) running automated processes in an ArcGIS environment, using a custom Python script that implements a series of ArcGIS functions to generate the network features, and 2) manually editing the initial network for quality control (which is unavoidable due to inconsistencies among data types). Figure 1 shows the steps for creating the sidewalk network database.

The research team selected Midtown Atlanta to illustrate the methodology, given the diversity of roadway geometries and land use densities (*11*). This mixed-use Atlanta neighborhood includes approximately 10,000 residential households (evenly split between single-family and multi-family housing), and 800 high-density commercial parcels that support about 65,000 daytime workers across a spectrum of job types. The main steps of the methodology are described below.

Data Inputs

Primary inputs are parcel-level data and roadway centerline data. Parcel boundary GIS data are typically comprised of polygons representing the property plats within a jurisdiction. Parcel boundaries usually represent the boundaries of the public right-of-way and, at least initially, represent the expected approximate sidewalk locations along the street

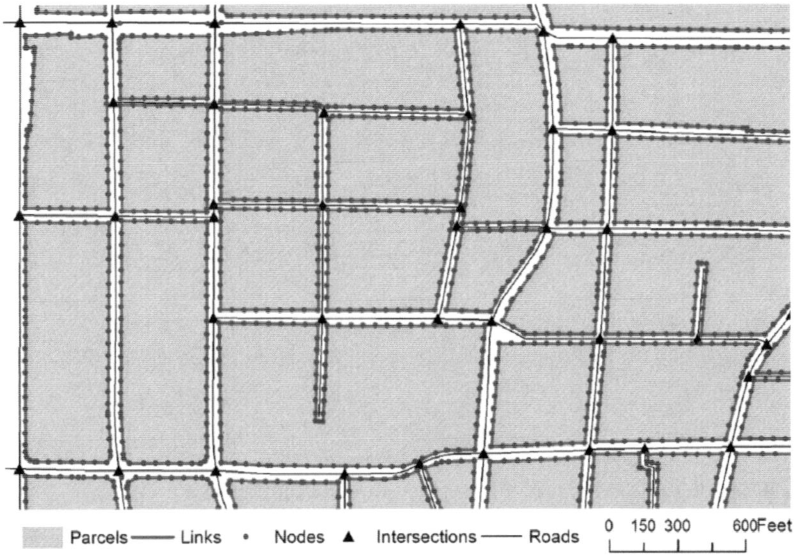

Figure 2. Generation of sidewalk links and nodes.

frontage. There are several reasons that the research team chose these data sources as network generation inputs. Most state and local transportation departments maintain a roadway centerline file for transportation network modeling and roadway asset management. Many jurisdictions maintain high-quality parcel-level GIS data to manage property transfers, building permits, and property tax assessment. Parcel-level land-use data and roadway centerline data are often available to the public at no cost. In addition, these data sources are typically accurate and frequently updated.

Generate Intersections

Roadway centerlines are clipped to the analysis boundaries and visually checked for any major discrepancies prior to beginning the analysis. Verification can be easily achieved using a background map embedded in ArcGIS. Intersection nodes are then generated using roadway centerlines as follows: 1) generate breaking points of road segments; 2) intersect breaking points with road segments and filter out the points that intersect with fewer than three legs; and 3) label each intersection with a unique ID and add an attribute column containing the number of legs per intersection.

Generate Sidewalk Links and Nodes

Parcel polygons are aggregated at the block level, and then polygons are simplified by removing points within 5 ft to achieve a proper net detail. Manual editing is likely required to identify and correct aggregation errors. However, corrections are not performed until after aggregation because: 1) some typology errors (e.g., parcel overlap) are automatically eliminated during aggregation, and 2) quality assurance/quality control (QA/QC) after aggregation would need to be performed anyway, even if underlying parcels were corrected, because some parcels may not aggregate properly. After parcel aggregation and verification, sidewalk links are generated using the sides of the aggregated parcel polygons. In essence, these are the expected locations of the sidewalk network. The links are then broken into approximately 50-ft link segments and assigned unique identifiers for use in asset management work. Through testing, the standard 50-ft length was determined to be a good length for summarizing sidewalk data that is joined to the network and used for planning purposes. Sidewalk links vertices are coded as sidewalk nodes, which are also potentially ramp nodes when located near roadway intersections.

Figure 2 illustrates the sidewalk link and node network in Midtown Atlanta, which uses a 25-ft aggregation threshold (selected after conducting several tests in the dense Atlanta neighborhoods of Virginia Highland, Home Park, and Candler Park). At this stage, sidewalk links surround each parcel, but still need to be connected to sidewalks surrounding adjacent parcels via crosswalks at intersections to reflect real-world walkability.

Generate Crosswalk Links

Although it is relatively simple to manually edit the developed sidewalk network and add the crosswalk links for a small area, another purpose of the research was to automate this step and facilitate sidewalk GIS network generation for metropolitan areas. First, sidewalk nodes within a 60-ft buffer of each intersection are identified and named as ramp

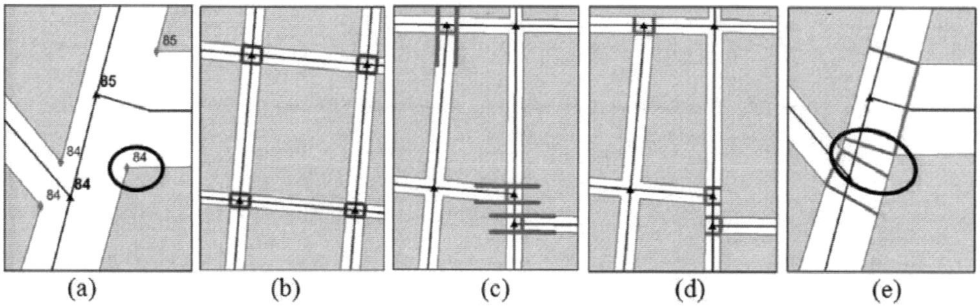

Figure 3. Generation of crosswalk links. (*a*) Extra identified nodes. (*b*) Four-leg intersection: all links. (*c*) T-intersection: link candidates. (*d*) T-intersection: all links. (*e*) Closely aligned intersections.

nodes. The spatial selection threshold can vary in different datasets, but the goal is to ensure that the majority of ramp location nodes are automatically included. A manual check is necessary to ensure that ramp nodes were properly selected and properly labeled with the applicable intersection ID. For example, in Figure 3*a*, one node at Intersection 85 was incorrectly assigned to Intersection 84 because it is closer to that intersection, and is manually revised to Intersection 85. After verification, scripts generate crosswalk links with two different methods, one for intersections with four or more legs, and one for intersections with three legs (T-intersections).

For a typical four-leg intersection, four ramp nodes are selected and four of the six possible combinations of two nodes serve as crosswalk link candidates. As the ramp nodes form a quadrilateral, the four boundary sides are automatically selected to generate crosswalk links, illustrated in Figure 3*b*. Diagonal crossings are rare, and can be added manually if necessary.

For a T-intersection, generating crosswalk links is challenging because there are no nodes along the edge of the opposing parcel. Therefore, one crosswalk per T-intersection is created first (named first crosswalk), directly connecting the two nodes at the intersection using the node coordinates (see red links in Figure 3*c*). Two additional crosswalk links are created, based on the assumption that these two links are parallel with each other, and perpendicular to the first crosswalk. Because there is no readily available information about the length of these two crosswalks or their bearings (two opposite directions are both perpendicular to the first crosswalk), four 100-ft line features per intersection are automatically generated as crosswalk link candidates. To extract correct crosswalk links, the script clips these link candidates using the boundaries of aggregated parcels and then filters out erroneous crosswalk links, illustrated in Figure 3, *c* and *d*.

QA/QC of the automatically generated crosswalk links is necessary at this point, given the diversity of intersection configurations and parcel data inaccuracies. Figure 3*e* illustrates an example in which one of the two links closest to each other is discarded to represent a single crosswalk. Manual labor for crosswalk link editing is lower when nodes of parcel corners are properly selected and labeled.

Create a Fully Connected Sidewalk Network

A fully connected sidewalk network consists of links (sidewalk and crosswalk), with each link joining two nodes. Because T-intersection crosswalk links are not generated from sidewalk nodes directly, sidewalk links are split by these nodes. A visual quality check is conducted to ensure the connectivity of the network, and then the links and nodes are stored in a sidewalk network geodatabase. Figure 4 illustrates the sidewalk network for Midtown Atlanta. This network is readily available to load inventory data, that is, crosswalk links, sidewalk links, ramp nodes and sidewalk nodes can carry crosswalk, sidewalk, ramp, and curb ramp/cut design and condition data.

Evaluation of the Proposed Method

To assess the efforts and effectiveness of the automated method, the research team applied the method to six representative neighborhoods in the City of Atlanta, including Midtown, Home Park, Ansley Park, Virginia Highland, Morningside, and Candler Park. All of the automated steps were successful for these neighborhoods, generating sidewalk links and nodes as expected in just a few minutes of computer processing. Table 1 shows the results of a connected sidewalk network. The combined network includes 245 mi of sidewalk links, 6,432 sidewalk nodes, 760 intersection nodes, and 2,120 crosswalk links.

As noted throughout the methodological section, the semi-automated method also requires manual QA/QC analysis. The team summarized the results of the manual QA/QC analysis in the metric of labor hours consumed in the process in Table 2. For each neighborhood, loading data and implementing the script only took about 10 min on an Intel i7-2600k CPU 3.4 GHz computer. The time spent in manual QA/QC processes varied between 1.0 h and 2.4 h per neighborhood, with an average of 1.3 h/mi^2. Correcting intersections and

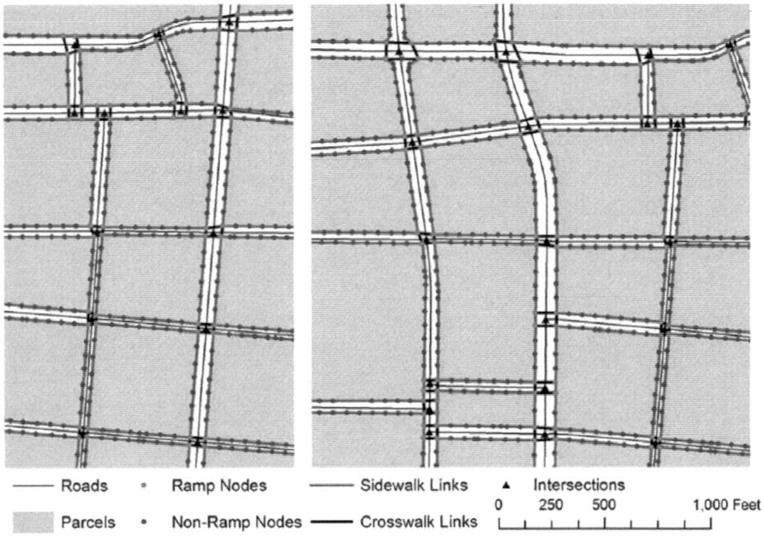

Figure 4. Midtown sidewalk network.

Table 1. Processed Intersections, Parcels, Sidewalk Links, and Nodes

Neighborhood	Land area (mi^2)	Parcel area (mi^2)	Link length (mi)	Number of features				
				Intersections	Parcels	Sidewalk nodes	Sidewalk links	Crosswalk links
Midtown	1.6	1.3	64.4	216	160	1,363	1,358	660
Home Park	0.7	0.5	26.3	97	62	545	545	268
Ansley Park	0.6	0.6	22.6	77	43	914	914	198
Virginia Highland	1.0	0.9	40.9	119	69	964	964	329
Morningside	2.3	1.9	68.1	182	100	2,234	1,784	492
Candler Park	0.7	0.6	22.9	69	36	412	412	173
Total	6.9	5.8	245.3	760	470	6,432	5,977	2,120

aggregated parcels constitutes the majority of the time required for manual editing. The highest percentage of manual edits is in the intersection identification and crosswalk generation step. Because the correct selection of nodes directly influences the correct generation of crosswalk links, greater manual editing of intersection data on the front-end leads to less editing of the crosswalk links in subsequent steps. Overall, the study shows that the method developed performs significantly better at grid-type street network with standard intersection configurations.

Assuming that the City of Atlanta is 134 mi^2, with a population of 473,000, and GIS verification labor runs US$25/h, the total cost of generating Atlanta's sidewalk network costs less that one cent per person.

Application of the Sidewalk Network

To showcase the potential capabilities of the sidewalk network system presented in this study, this chapter presents sidewalk network data from Northampton, Massachusetts (total area: 35.8 mi^2) and East Point, Georgia (total area: 14.69 mi^2). These examples show the potential utility of sidewalk networks loaded with quantitative sidewalk condition data for establishing priorities for sidewalk repair and accessible pedestrian routing analysis. The following section is organized as follows: (1) a description of the systems used to collect the data in both communities, (2) an explanation of how collected data are tied to the sidewalk network, and (3) a simple demonstration of the pedestrian routing capabilities of the network.

Sidewalk Data Collection System

Sidewalk inventory data (location, design, and quality) were collected in Northampton in 2016, using data collection systems developed by researchers from Georgia Tech. The team developed a series of web- and smartphone-/tablet-based tools and methods to automatically collect and assess sidewalk

Table 2. Time Estimation for Generating and Processing the Sidewalk Network

	Generation and QAQC Time (min)					Labor
Neighborhood	Intersection	Sidewalk links and nodes	Crosswalk links	Time buffer	Total	(h/mi²)
Midtown	19	52	47	10	128	1.3
Home Park	16	30	28	10	84	2
Ansley Park	9	23	29	10	71	1.9
Virginia Highland	22	24	24	10	80	1.3
Morningside	17	69	49	10	145	1.1
Candler Park	12	29	11	10	62	1.6
Total	95	227	188	10	520	1.3

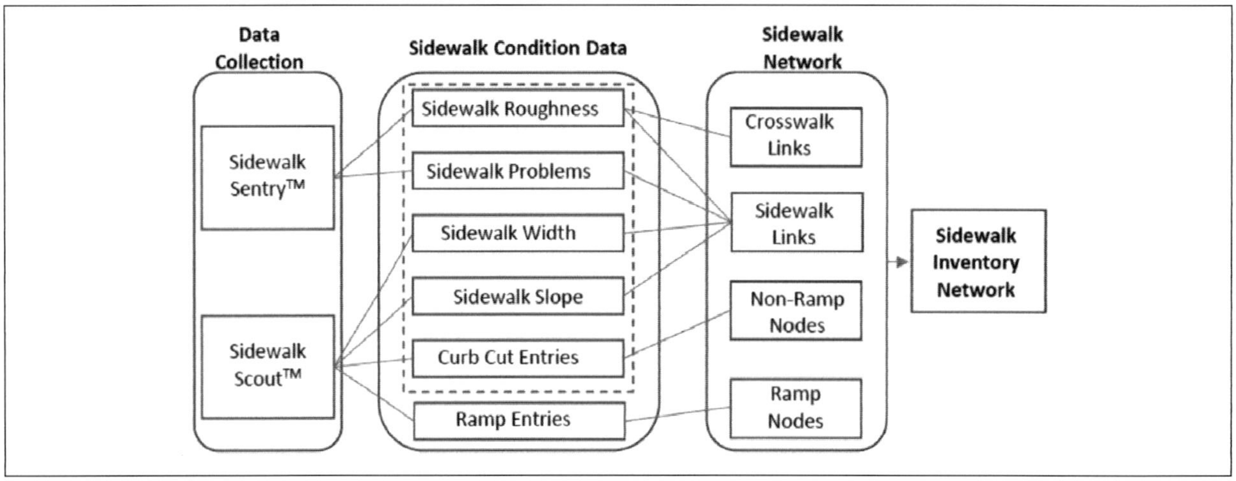

Figure 5. Flow chart of loading inventory data to the sidewalk network.

quality on a neighborhood or citywide scale at a low cost (*12*). Sidewalk inventory data were collected using Android- and IOS-compatible smartphone and tablet applications. Sidewalk Sentry™ is used for collecting sidewalk rolling video and vibration data to generate sidewalk roughness and defect reports (*13*), and Sidewalk Scout™ supports the collection of sidewalk slope, width, and curb ramp attributes reported by trained users and the general public that can be compared with ADA design standards (*14*).

Load Sidewalk Condition Data to the Sidewalk Network

Using the proposed method, the research team created a fully connected sidewalk network for Northampton, consisting of 2,679 sidewalk links, 2,019 crosswalk links, and 4,038 ramp nodes. Likewise, the East Point network consisted of 6,618 sidewalk links, 3,231 crosswalk links, and 3,384 ramp nodes. Sidewalk condition data were collected using Sidewalk Sentry™ and Sidewalk Scout™ and linked to the sidewalk links and node identifiers to generate a loaded sidewalk inventory network (Figure 5).

The collected sidewalk inventory data are linked with the sidewalk network using simple spatial join and snapping tools in GIS. Data are summarized by 50-ft sidewalk links, which are linked to an average sidewalk roughness score per link, a count of the problem reports by type, a count of non-compliant curb cuts within the segment, and an identifier as to whether the slope and width of the sidewalk segment are ADA compliant. Ramp data are similarly joined to ramp nodes, and intersection nodes are classified by whether they have no ramps, have non-ADA-compliant ramps, or have ADA-compliant ramps. Intersection nodes are also summarized by the number of ADA-compliance issues. Crosswalk links are classified as having no ramps on either side, non-complaint ramps, or compliant ramps for network routing purposes. In the future, the team plans to include additional ADA-compliance data points such as pedestrian signals and crosswalk design. Figure 6 shows examples of the summarized data from Northampton after it was processed. These maps show ADA-compliance issues that are typically present in communities that have not been inventoried or assessed in the 25+ years since the ADA was passed.

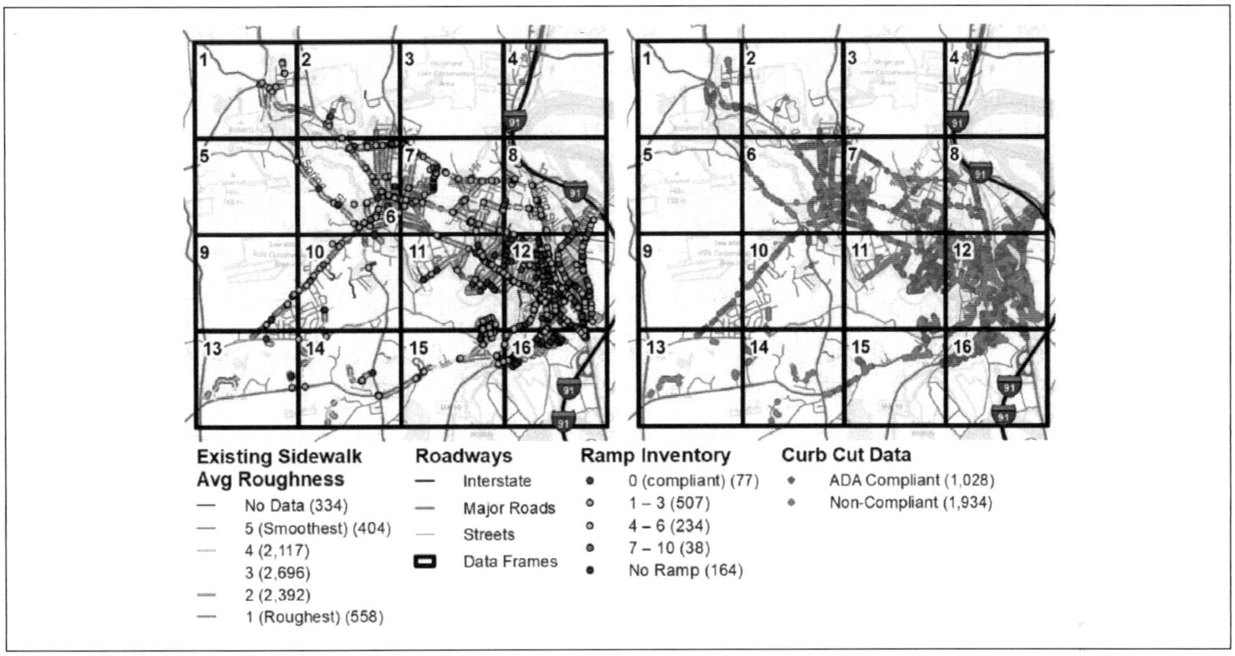

Figure 6. Sidewalk inventory network in Northhampton.

Pedestrian Routing

The sidewalk inventory network also allows for pedestrian routing in GIS. Here, we show a simple origin–destination pair scenario in East Point to assess the impact, if any, that sidewalk deficiencies have on travel time and route choice. An origin and destination pair was chosen in downtown East Point, which has the greatest sidewalk coverage and consequently, the greatest amount of sidewalk quality data collected. ADA-compliance issues were assigned impedances on travel time (subjective estimates) with the more severe issues (such as missing ramps) receiving a higher travel-time impedance. These impedance times were added to the estimated average walking times for each sidewalk and crosswalk segment (assuming a walking speed of 3.0 ft/s). Sidewalk problem reports added 1.0 s each to the total walk time, ramp ADA issues added 0.5 s each to the total walk time, missing curb ramps added 5.0 s to the total walk time, and sidewalk segment roughness score increments worse than "best" added 1.0 s per increment to the total walk time. The ArcMap Network Analyst extension was utilized to determine the shortest route, with regard to time, between the two points. Comparing the impedance weighted route to the non-weighted route (Figure 7) shows a potential difference in preferred route when ADA issues are considered. Testing and using actual impedance times that best reflect ADA issues (missing ramps, pavement disjoints, potholes, etc.) could make this routing capability more accurate and useful to those with disabilities in the future.

Conclusions

The effective implementation of sidewalk asset management systems requires a linkage between high-resolution sidewalk quality data collected in the field and structured network features. Similar to a roadway network in a GIS environment, which enables storing, visualizing, and analyzing quantitative and qualitative roadway characteristics, a GIS-based sidewalk network is essential to sidewalk infrastructure management. Hence, developing independent sidewalk networks will be essential for comprehensive sidewalk network evaluation and for making informed pedestrian infrastructure improvement decisions. This study used parcel and roadway data to develop sidewalk network databases through a semi-automated process utilizing built-in ArcGIS tools, scripting in the ArcGIS environment, and manual editing during QA/QC. The algorithms generate sidewalk and crosswalk links, and ramp nodes that should exist in theory. Missing sidewalks can be coded with width = 0 and missing ramps can be coded as "absent."

To assess the efforts and effectiveness of the automated method in creating the sidewalk network, the research team quantified the time spent in manual QA/QC and data editing processes per linear mile and land area (square-miles) for six representative neighborhoods in the City of Atlanta. Based upon these empirical tests, the average time involved in manual editing was about 1.3 h/mi^2 (2 h was the maximum time spent to finalize a neighborhood). Labor cost is expected to be significantly lower when high-quality input data are available, which is promising as more and more states are developing their own high-quality roadway and parcel data.

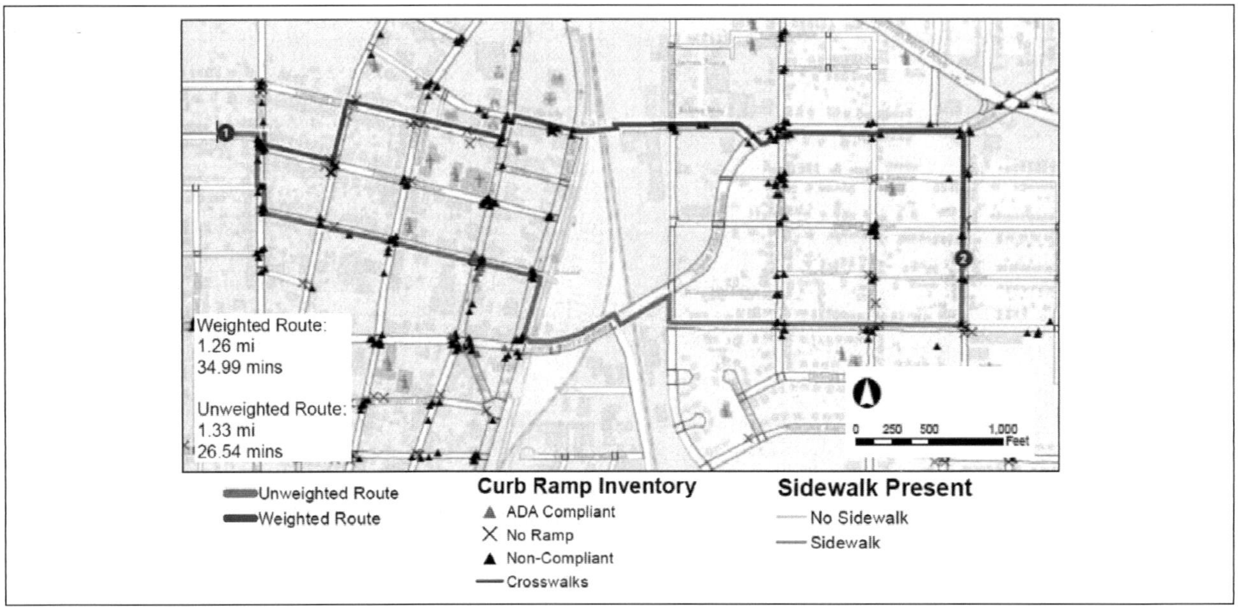

Figure 7. Pedestrian routing based on sidewalk inventory network.

Compared with prior studies that used GPS traces and/or aerial images, a major advantage of the proposed method is the low labor cost and scripting efficiency. Good quality parcel data and roadway data are also available in many regions, free of charge. The sidewalk network generated by the process is fully connected, well structured, and readily available to carry sidewalk condition data. Manual correction is unavoidable due to the data quality and complexity of certain intersections. An inventory of irregularities in the sidewalk network can also significantly reduce manual efforts.

To date, several cities, including Yarmouth (Maine), Northampton (Massachusetts), East Point (Georgia), and Atlanta (Georgia) have implemented the proposed method. The authors are in the process of creating the sidewalk networks for the entire Atlanta metro area. For large area QA/QC assessment, more advanced techniques will be developed and applied. Once city managers have access to an underlying sidewalk network populated with sidewalk, ramp design, and condition data, planners and decision-makers will be able to better visualize, analyze, and evaluate the pedestrian infrastructure, and efficiently plan and prioritize infrastructure repair and construction decisions.

Acknowledgments

Various elements of the work presented in this paper were supported by, in part, the U.S. Department of Transportation, Georgia Department of Transportation, and the Georgia Institute of Technology.

References

1. Landis, B., V. Vattikuti, R. Ottenberg, D. McLeod, and M. Guttenplan. Modeling the Roadside Walking Environment: Pedestrian Level of Service. *Transportation Research Record: Journal of the Transportation Research Board*, 2001. 1773: 82–88.
2. Woodcock, J., P. Edwards, C. Tonne, B. G. Armstrong, O. Ashiru, D. Banister, S. Beevers, Z. Chalabi, Z. Chowdhury, A. Cohen, and O. H. Franco. Public Health Benefits of Strategies to Reduce Greenhouse-gas Emissions: Urban Land Transport. *The Lancet*, Vol. 374, No. 9705, 2009, pp. 1930–1943. http://dx.doi.org/10.1016/S0140-6736(09)61714-1
3. Bodea, T. D., L. A. Garrow, M. D. Meyer, and C. L. Ross. Explaining Obesity with Urban Form: A Cautionary Tale. *Transportation*, Vol. 35, No. 2, 2008, pp. 179–199. http://dx.doi.org/10.1007/s11116-007-9148-2.
4. Barden, v. City of Sacramento. 292 F. 3d 1073-Court of Appeals, 9th Circuit, 2002. http://openjurist.org/292/f3d/1073/barden. Accessed September 29, 2016.
5. Karimi, H. A., and P. Kasemsuppakorn. Pedestrian Network Map Generation Approaches and Recommendation. *International Journal of Geographical Information Science*, Vol. 27, No. 5, 2013, pp. 947–962. http://dx.doi.org/10.1080/13658816.2012.730148.
6. Ballester, M. G., M. R. Pérez, and J. Stuiver. Automatic Pedestrian Network Generation. *Proc., 14th AGILE International Conference on GIS*, Utrecht, the Netherlands, 2011, pp. 1–13.
7. Kim, J., S. Park, Y. Bang, and K. Yu. Automatic Derivation of a Pedestrian Network Based on Existing Spatial Data Sets. *Proc., ASPRS/MAPPS 2009 Fall Conference*, San Antonio, TX, USA, 2009, pp. 16–19.
8. Parker, J. S., and E. Vanderslice. Pedestrian Network Analysis. In *Walk21 Vancouver, Reducing Road Dangers, Walk* 21, 2011.
9. van Eggermond, M. A., and A. Erath. Pedestrian and Transit Accessibility on a Micro Level: Results and Challenges. *Journal of Transport and Land Use*, Vol. 9, No. 3, 2016, pp. 127–143. http://dx.doi.org/10.5198/jtlu.2015.677

10. Khoeini, S., V. Elango, Y. Xu, and R. Guensler. Generating Sidewalk Network Using Roadway Network and Parcel Data. Presented at 94th Annual Meeting of the Transportation Research Board, Washington, D.C., 2015.
11. Midtown Alliance. Midtown Alliance Resource Center. http://www.midtownatl.com. Accessed September 29, 2016.
12. Guensler, R., A. Frackelton, A. Grossman, V. Elango, Y. Xu, C. Toth, A. Akanser, F. Castrillon, S. Khoeini, E. Palinginis, and R. Sadana. *Automated Sidewalk Quality and Safety Assessment System*. Prepared for the Georgia Department of Transportation and STRIDE University Transportation Center. Georgia Institute of Technology, Atlanta, Ga., 2015.
13. Frackelton, A., A. Grossman, E. Palinginis, F. Castrillon, V. Elango, and R. Guensler. Measuring Walkability: Development of an Automated Sidewalk Quality Assessment Tool. *Suburban Sustainability*, Vol. 1, No. 1, 2013. http://dx.doi.org/10.5038/2164-0866.1.1.4
14. Akanser, A., V. Elango, A. Grossman, Y. Xu, and R. Guensler. *Sidewalk Scout: Crowdsourcing Android Application for Collecting Sidewalk Condition Data*. Transportation Research Forum, Atlanta, Ga., 2015.

The Standing Committee on Transportation Asset Management (ABC40) peer-reviewed this paper (18-02656).

Article

TRR
JOURNAL OF THE TRANSPORTATION RESEARCH BOARD

Data Fusion of Commercial Vehicle GPS and Roadside Intercept Survey Data

Transportation Research Record
2018, Vol. 2672(44) 10–20

DOI: 10.1177/0361198118768516
journals.sagepub.com/home/trr
SAGE

Sirui Zhu[1], Glareh Amirjamshidi[2], and Matthew J. Roorda[1]

Abstract

GPS tracking technology produces large amounts of data which represent samples of the commercial vehicle population that are much larger than conventional commercial travel surveys. However, passively collected GPS data lack behavioral detail that a conventional survey offers. This study develops a data fusion method to impute variables of interest for a large GPS data set, by establishing a link to a behaviorally rich commercial travel survey data set. As a case study, this study uses detailed information from the Ministry of Transportation of Ontario's Commercial Vehicle Survey (CVS), a truck intercept survey conducted in 2012, to enrich a GPS commercial vehicle tracking data set from Xata Turnpike Inc. The enrichment process has three parts: converting raw GPS tracking data into GPS trips, matching CVS trips to GPS trips, and imputing the missing variables for GPS trips. Evaluation of the outcomes concludes that imputation methods can produce a synthetic data set with large sample size (from GPS data) and rich information (from roadside interview data) with good accuracy.

GPS tracking technology produces large amounts of data which represent samples of the commercial vehicle population that are much larger than conventional commercial travel surveys. Unlike conventional surveys, which typically take a cross-sectional "snapshot" of commercial operations, these passively collected data can be collected continuously over large geographic areas over long periods of time, with little burden on the operator. Such continuous data collected through GPS can enable public sector analysis and modeling of commercial travel with great spatial and temporal detail, and longitudinal aspects such as seasonal variation. However, passive technology-based data collection sacrifices some of the behavioral richness offered in survey-based techniques, most importantly the collection of data about the vehicle, driver, stop purpose and cargo (commodity being carried and whether the vehicle is loaded or empty). Data fusion techniques to integrate information from large passively collected GPS data sets with richer behavioral data from commercial travel surveys would clearly be of great benefit in this context.

Data fusion is the process of integrating data from multiple sources to produce more meaningful information. Data fusion has recently appeared in the transportation literature, but has been applied in other fields for a considerably longer time (*1*). Bachmann et al. compared several multi-sensor data fusion methods for estimating freeway traffic speed using loop detector and probe vehicle data (*2*).

The objective of this research is to develop a customized data fusion method to integrate two data sets available to the Ministry of Transportation of Ontario (MTO): Commercial Vehicle Survey (CVS) and GPS data acquired from a company specializing in commercial fleet management. The output of the data fusion effort in this study is a synthetic data set that encompasses all truck tours tracked by GPS data, augmented with synthetically generated behavioral attributes (e.g., commodity type and vehicle type) for each truck trip, reflecting data from the CVS.

The CVS is one of MTO's main instruments for obtaining information about freight vehicle flows on the provincial transportation system. The CVS is a roadside survey that intercepts truck trips at data collection sites (DCS) to gather information about inter-city truck flows as well as some urban truck flows (*3*). GPS tracking data comprise detailed positions (latitude and longitude) with time stamps for each individual truck at frequent time intervals. Connecting the position data in chronological order yields a detailed representation of the tour for each truck. However, the device does not collect data with more complexity such as cargo and vehicle type. Relative to the cost of the CVS, the purchase of GPS tracking data is, however, inexpensive.

This research is undertaken to enhance potential freight modeling and analysis capabilities of the MTO. The resulting enriched database would provide much greater support for commodity-based or tour-based models of commercial vehicle operations in Ontario. It would also fill in some gaps that are present in the representation of freight flows in Ontario becuase of the limited interview locations of the CVS.

[1]Department of Civil Engineering, University of Toronto, Toronto, ON, Canada
[2]DeGroote School of Business, McMaster University, Hamilton, ON, Canada

Corresponding Author:
Address correspondence to Glareh Amirjamshidi:
glareh.amirjamshidi@alum.utoronto.ca

Finally, it would allow for a more continuous representation of freight flow variations by time of year, as a continuous stream of GPS data could be purchased by the MTO.

The next section provides an overview of data fusion methods. Data available for this study are then described. The following section describes the methods for data preprocessing and imputation, followed by results and recommendations for future research.

Background

Data Fusion Methods

Data fusion is the process of synthesizing data from multiple sources to produce more meaningful information. To integrate two or more data sets in a data fusion procedure, the data sets must have the following characteristics (*4*):

1. There must be a set of variables that are common across all data sets (common variables);
2. Each data set must also have its own set of variables that are not jointly observed by another data set (specific variables);
3. The units observed in each data set are different units from another data set.

There are two procedures for data fusion: exact matching and statistical matching. Exact matching is a process in which an individual from a data set is matched with the same individual from a different data set. In other words, each record is unique and identifiable, and it is only matched with a record for the same individual when linking data between different data sets (*5*). Exact matching often uses identifiers such as name, address, or social insurance number to match the same subject across different data files. Studies on exact matching mainly focus on identifying causes of matching errors and optimizing the computer matching process (*6*).

Statistical matching attempts to match records from different data sets based on their similarity in common characteristics rather than unique identification information (*6*). A record in one data set can be matched to multiple records in a different data set, as long as the common variables are sufficiently similar. The data set from which specific variables are copied to the other data set is called the donor data set. The data set with no information for the specific variables, which accepts data from the matched donor record, is called the receptor data set. This study applies statistical matching because no identifying information is provided in either the CVS data or the GPS data.

Statistical Matching

In statistical matching for data fusion, there are two main approaches: explicit modeling and implicit modeling. The explicit modeling approach (also called the classical approach) develops models, such as multiple regression, logistic regression, or tree-based models, such that specific variables are the outputs of a function of the common variables in the donor data set and a vector of parameters. These models are then applied to the receptor data set to produce results for records that are missing information on specific variables (*7*). The main advantage of explicit modeling is that the process relies on a statistical framework with minimal subjective decisions made by the modeler. Another advantage is that data for records with little similarity can be imputed as long as the joint distribution is established, although the statistical accuracy may be unsatisfactory. However, if there is no prior knowledge about the possible relationship between variables, the parameters may be difficult to estimate. Explicit modeling is also computationally intensive because it needs to conduct multiple imputations using many multidimensional distributions. More information on explicit modeling is beyond the scope of this paper but can be found in (*1, 2, 4*).

Implicit modeling transfers values of specific variables based on records with no missing data and similar common variables (*5*). Implicit modeling relies on similarities between observed data sets to transfer values from donor records to receptor records. Therefore, implicit modeling can be considered a special case of data imputation. Data imputation refers to the process of populating missing data in a data set by using similar records in the same data set (*8*). For the case of data fusion, the process involves populating missing data in the receptor data set using similar records from the donor data set.

The most well-known imputation modeling technique is hot-deck imputation, which selects a donor record that shares the closest common variables with the receptor record and transfers the specific variables from that donor record to the receptor record (*9*). The definition of "closest" refers to the difference in common variable values between the two records, and the method to calculate the difference can vary. Many variations and improvements have been made to the hot-deck imputation technique, such as random hot-deck, sequential hot-deck, and nearest neighbor donor (*8*).

For the techniques of selecting a pool of donor records (i.e., a sample on which the imputation of the receptor will be based), there are numerous papers that provide computation algorithms, each with advantages and disadvantages. D'Ambrosio et al. (*7*) present a tree-based incremental imputation method, which uses a decision-tree structure to sort the specific variable of interest in sequence from variables with the least missing data to those with the most missing data. The process then imputes the first specific variable with missing data, which would have the highest number of records with actual values available in the specific variable, for the best probability of finding a similar record. The process is repeated iteratively until all missing data are imputed. This method excels at the fusion of list-based data sets and minimizes the number of probability distributions to be computed (*7*).

The implicit modeling approach can use constrained or unconstrained imputation. Unconstrained imputation is the process by which a donor record can be assigned to any number of receptor records, as long as that donor record meets the requirements of donor pool selection for implicit modeling. Constrained imputation, on the other hand, assigns weights to the records in the receptor data set and donor data set that must be preserved after matching. The weight limits the contribution of a donor record to a receptor record. The constrained imputation avoids the bias of certain donor records being overrepresented by limiting the contribution a single donor record can make (*10*).

The main advantage of the implicit modeling approach is its ease of computation. It primarily uses searching and matching functions rather than optimization functions with multidimensional distributions. In addition, the implicit modeling approach requires less understanding of the correlation between the variables in the data set. This is particularly useful for this study, in which there are a limited number of highly informative common variables between the CVS and GPS data sets. Given the limitation of data availability and the ease of transferability of a method to other forms of GPS probe data, this study focuses on the implicit approach.

Data

Site Selection

The Trafalgar DCS are located at eastbound and westbound weigh-scale stations on Highway 401 near the Trafalgar Road Interchange close to the Town of Milton, Canada just west of Toronto. These sites were selected for this study by the Ministry of Transportation because (1) it is one of the busiest sections of highway in Canada (annual average daily traffic of 138,500), and (2) in the CVS data, there are 3,795 records collected at the Trafalgar DCS and 6,377 records taken from other DCS that contain vehicle movements that passed through the Trafalgar DCS en route. This makes a total of 10,172 CVS records (approximately 25%) to represent the freight flow through Highway 401 at Trafalgar Road.

CVS Data Set

The CVS, a roadside interview of truck drivers, is conducted approximately every 5 years, and the most recently available data were collected in 2012. The 2012 CVS data set consists of 45,192 records over 200 DCS, with each record representing an interview with a truck driver at a DCS in Ontario. The CVS interview asks the driver to fill out an extensive survey questionnaire that encompasses information about the vehicle configuration, driver characteristics, commodity attributes, delivery addresses, and more. The survey questions shed light on the cross-sectional status of freight vehicles at that specific time and location, but do not give details about trip information such as departure/arrival time, trip route, and duration of stay at stops before or after the survey. Although the survey does ask for trip origin and trip destination, no details are requested about intermediate stops except for the previous stop immediately before the survey and next stop immediately after the survey. The immediate previous stop and next stop can be any planned stop by the driver such as service stop, stopping at home, reporting to head office, and so on; they are not limited to loading/unloading stops. However, the survey does not collect information about the purpose of the previous stop and next stop, so one cannot be certain if the intermediate stops are loading stops. Although the CVS is an extensive data collection effort, the sample collected through the CVS is still a small portion of the entire truck population that travels in Ontario (*3*).

GPS Truck Data

The GPS truck data provided by MTO consists of 1,881 data files collected by Xata Turnpike (now Omnitracs), a firm that provides GPS tracking and fleet management services for many commercial fleet operators. Each file contains the movement of a single freight vehicle over the course of October 2011 that passed through the Trafalgar DCS at least once (*3*). Each file contains GPS points, the number rangi from 3,000 to over 8,000. The movement of the vehicle is represented by the series of GPS points, showing the latitude and longitude coordinates, vehicle instantaneous speed, and time stamp. Through pattern observation of the GPS data, it was noticed that the GPS device checks the location and time of a vehicle at a 5-s interval but only stores this information after the vehicle has travelled a distance of 500 m from the previously recorded location if the speed is below 80 km/h. When the vehicle reaches speeds that are only feasible on freeways, GPS points are stored after travelling for 1 mi. For situations in which the vehicle is travelling slowly, a GPS point is also stored when more than 80 s passes and the distance from the previously recorded point is less than 500 m. This is further discussed in the paper by McCabe et al., which uses similar GPS data from the same service provider Xata Turnpike (*11*).

Method

The data imputation process has three main steps which are further discussed in this section.

1. Clean and preprocess the GPS data into a trip format that is comparable with CVS data;
2. Identify matching CVS candidate records for each GPS trip and quantify the quality of each match, based on the "spatial scale" and the "stop combination";
3. Select a CVS candidate (based on the quality of the match) and assign its attribute values to the GPS trip.

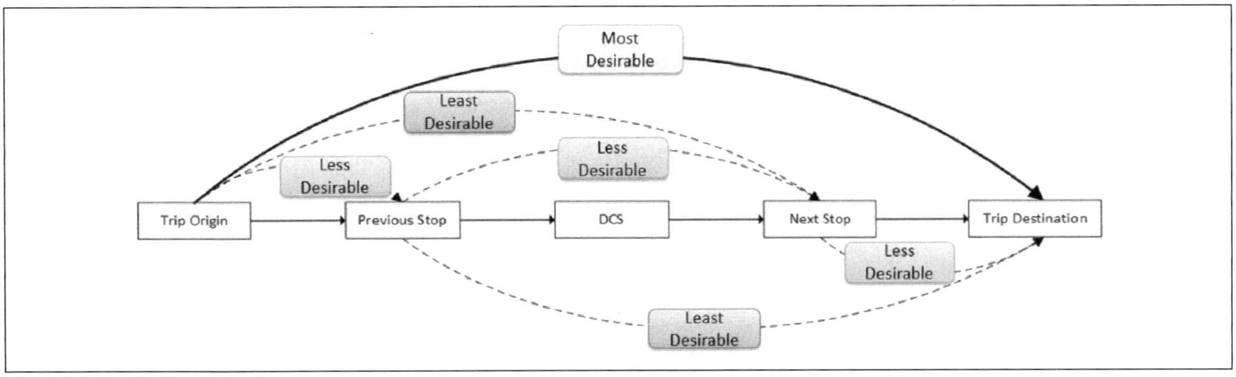

Figure 1. Possible stop combinations and desirability of each combination for matching.

Clean and Preprocess GPS Data

GPS points that are of interest in this research are points where the vehicle has stopped moving to load or unload cargo. Therefore, the first step for the GPS data set filters and identifies GPS points that are potential trip ends. This study uses standards similar to those applied by Sharman et al., who used similar data from Turnpike Route Tracker™ GPS units to model activity behavior and stop duration of urban commercial vehicles (*12*). The steps are as follows.

- Filter 1: Remove stops less than 5 min in duration to reflect the 5-min time threshold for recording stops set by Turnpike for the RouteTracker™ unit.
- Filter 2: Remove stops within 40 m of freeway centerlines.
- Filter 3: Remove stops within 20 m of major arterial road centerlines.
- Filter 4: Remove stops at USA/Canada border custom inspection areas.
- Filter 5: Remove stops at Ontario highway rest stops.
- Merge stops within a single business establishment.
- Remove points that represent unrealistic 'jumps' in space and time.

Consecutive GPS trip ends identified through the filters above define GPS trips that are used to match with trips in CVS data. More detail about the data cleaning and preprocessing can be found in (*12, 13*).

Identify Matching CVS Candidate Records for Each GPS Trip

In this stage, GPS trip origins and destinations are compared with CVS records, to check if they match. Matching is assessed using two criteria, spatial scale and stop type combination. A CVS record is considered a candidate match for the GPS trip if the GPS trip ends are at the same locations as the trip ends in the CVS record. The thresholds for defining "the same location" are referred to as spatial scale. Within Canada, five spatial scales are considered: exact coordinates within 200 m, Census Dissemination Area (DA), Census Subdivision (CSD), Census Division (CD), and within North America (NA).

CVS records attempt to capture freight vehicle touring behavior by asking about non-traffic stops immediately before and planned stops after the interview, known as previous stop and next stop. Most often, the previous stop and next stop are the same as trip origin and trip destination. However, there are some cases in which the previous stop and trip origin are different or the next stop and trip destination are different. Figure 1 shows all stop combinations that are possible in the CVS data, with some stop combinations more desirable for matching than others. The most desirable combination is the conventional trip origin to trip destination, because these stops are loading/unloading stops. Less desirable combinations are those for which the stops are immediately adjacent to each other chronologically. A stop combination of CVS previous stop to CVS next stop, for example, is a combination in which the stops are immediately after each other with no skipped stop in between. Stop combinations such as CVS origin to CVS next stop are not consecutive (the next stop is between) and therefore are considered least desirable.

The GPS trips are each checked against 10,172 CVS records that passed the Trafalgar DCS. If CVS record trip ends and GPS trip ends are at matched locations, that CVS record is considered a candidate for a GPS trip. The trip origin and trip destination of the GPS trip can be matched to the CVS candidate at different spatial scales. For example, a GPS trip origin may be in the same DA as the CVS record's trip origin, but the GPS trip destination only shares a common CD as the CVS record's trip destination.

A candidate that is matched at a spatial level of the exact location with 200 m for both trip origin and trip destination provides a better-quality match than a candidate that is matched at a spatial level of CD. Therefore, a penalty score system is implemented to account for the quality of each matched candidate, based on the spatial scale and stop

Table 1. Penalty Score Calculation Models and Penalty Values for Individual Components

Penalty score calculation models		
Variation 1	*Penalty Score 1 = exp(Origin Scale Penalty)+exp(Destination Scale Penalty) +Stop Combination Penalty*	
Variation 2	*Penalty Score 2 = (Origin Scale Penalty)^4+(Destination Scale Penalty)^4 +Stop Combination Penalty*	
Variation 3	*Penalty Score 3 = exp(Origin Scale Penalty)+exp(Destination Scale Penalty) +exp(Stop Combination Penalty)*	
Penalty values for individual components		
Spatial scale		Origin/destination scale penalty
Exact location with 200 m threshold		1
Census dissemination area		2
Census subdivision/us county		3
Census division		4
Within North America only		10
Stop type combination (see Figure 1)		Stop combination penalty
Trip origin	Trip destination	0
Trip origin	Previous stop	1
Trip origin	Next stop	3
Previous stop	Next stop	1
Previous stop	Trip destination	3
Next stop	Trip destination	1

combination associated with each candidate. A candidate that uses aggregate spatial scales would incur a higher penalty score than a candidate with disaggregate spatial scales. Similarly, a candidate with less desirable stop combination would receive higher penalty score than candidates with desirable stop combinations. Although both factors play a role in the quality of a candidate, it is assumed that spatial scale is a more important factor than stop combination.

Three variations of the penalty score calculation were tested in this study and are shown in Table 1. Variation 1 calculates the penalty score by adding the exponential of spatial scales and the value for stop combinations. Variation 2 uses a power function to represent the increase of penalty score by spatial scale instead of an exponential function, while the weight of the stop combination remains the same. Variation 3 differs from Variation 1 by making the stop combination also increase exponentially in the penalty score calculation such that the least desirable stop combination would contribute more to the penalty score relative to the less aggregate spatial scales. Application of the penalty score calculation results in a pool of matched CVS candidate trip records each with a penalty score. To determine the candidate that is selected to impute the attributes, two factors are considered:

- Degree of similarity between GPS trip and CVS candidate, reflected by the penalty score;
- National Expansion Factor (NEF) of the CVS candidate, which refers to the CVS record's representativeness of total freight movements in Ontario. The NEF was calculated by MTO for each CVS record. A CVS candidate with higher NEF is preferred because it represents a higher number of trips that passed the DCS.

To merge the two factors of quality and magnitude into a single measurement, a significance factor is defined as shown below. A high penalty score means poor quality match, and a high NEF implies a greater number of CVS trips are represented if this candidate is selected.

$$Significance\ Factor = \frac{NEF}{Penalty\ Score}$$

Select a CVS Candidate and Assign Its Attribute Values to the GPS Trip

Level of Aggregation of Attributes. Different attributes from CVS records should be applied to different groupings of GPS trip records, according to the nature of the attribute. Level of trip aggregation is the level of trip grouping at which the attribute is allowed one option for the entire group. Three levels of trip aggregation are assumed:

Table 2. Vehicle-based Attributes

Truck configuration attributes	1 Tractor and 1 trailer	5 Straight truck and trailer
	2 Tractor and 2 trailers	6 Tractor only (i.e., "Bob tail")
	3 Tractor and 3 trailers	95 Other
	4 Straight truck	99 Unknown (incomplete survey)
Body style attributes	1 Van—not refrigerated	14 Animal carrier
	2 Van—refrigerated	17 Utility (includes service vehicle)
	3 Container carrier	21 Concrete mixer
	4 Van—soft sided	24 Mobile machines
	5 Flatbed (with or without crane)	30 Piggyback
	6 Stake and rack	34 Towing equipment
	7 Float	35 Vehicle being towed
	8 Tanker (includes dry bulk carrier)	37 Bob tail only
	9 Hopper	38 Drop deck
	10 Dump	39 Logging truck
	11 Chip	94 Unknown (incomplete survey)
	12 Refuse carrier	95 Other (RV, combination body, carnival ride etc.)
	13 Vehicle carrier	
MTO commodity group codes	1 Agricultural products	11 Waste and scrap
	2 Food	12 Empty shipping container
	3 Minerals	13 Mail and parcel
	4 Petroleum and products	14 Mixed freight
	5 Chemicals and products	15 Equipment on company service
	6 Wood and products	16 Privately-owned goods being transported or serviced
	7 Metals and products	97 Unknown
	8 Machinery and electrical	98 Non-cargo carrying vehicle
	9 Manufactured products	
	10 Transportation	

- Vehicle-based: truck configuration, body style, and MTO general commodity group code (Table 2).
- Direction-based: detailed commodity type, that is, Standard Classification of Transported Goods (SCTG) division code (*14*), assumed to be the same for all trips by a truck in one direction.
- Trip-based: loading status (empty versus loaded), and cargo value which could vary from trip to trip.

Incremental Attribute Imputation Process. The imputation process of vehicle type attributes and the commodity attributes is outlined in Figure 2. The process starts by imputing vehicle-based attributes that are assumed to be common to all trips by the same truck. The imputed results are then used to narrow down the pool of candidates available for the imputation of loading status (empty or non-empty) and direction-based attributes. After further narrowing the pool for trips in each direction, trip-based attributes are then imputed. This process is similar to the tree-based incremental imputation method presented by D'Ambrosio et al. (*7*).

Attribute values are imputed as follows. First, all eligible CVS candidates of GPS trips within a trip group defined by the level of trip aggregation are pooled. Within this pool, each candidate is weighted based on the significance factor. Weighted random selection of a CVS candidate, with the weighting being by the significance factor, is then used as the imputation result. This approach is a hybrid of the simple hot-deck imputation method and the k-nearest neighbor (k-NN) imputation method. It is not a single hot-deck imputation because it does not select the "closest" candidate, which in this case is simply the candidate with the highest significance factor. It is not exactly the same as a k-NN method because there is not a predefined threshold "k" that qualifies a limited number of candidates and these candidates are not weighted equally. Once a candidate is selected and its attribute value is assigned to all GPS trips at the particular level of trip aggregation, the candidates in the pool are filtered to only allow candidates with the selected attribute value for subsequent attribute imputation at lower levels of trip aggregation.

In the example shown in Figure 2, the vehicle performed three trips. After selecting a single vehicle type (truck configuration and body style) for all three trips, the imputed results of the loading status step indicate that trips A and C are loaded whereas trip B is empty. All candidates for trip B are discarded, and a single MTO commodity group is assigned based on candidates for trips A and C. With that single MTO commodity group finalized, all candidates with SCTG divisions that are not a subset of the selected MTO commodity group are also discarded. The remaining candidates for trips A and C are used to impute SCTG division for the set of trips in each direction.

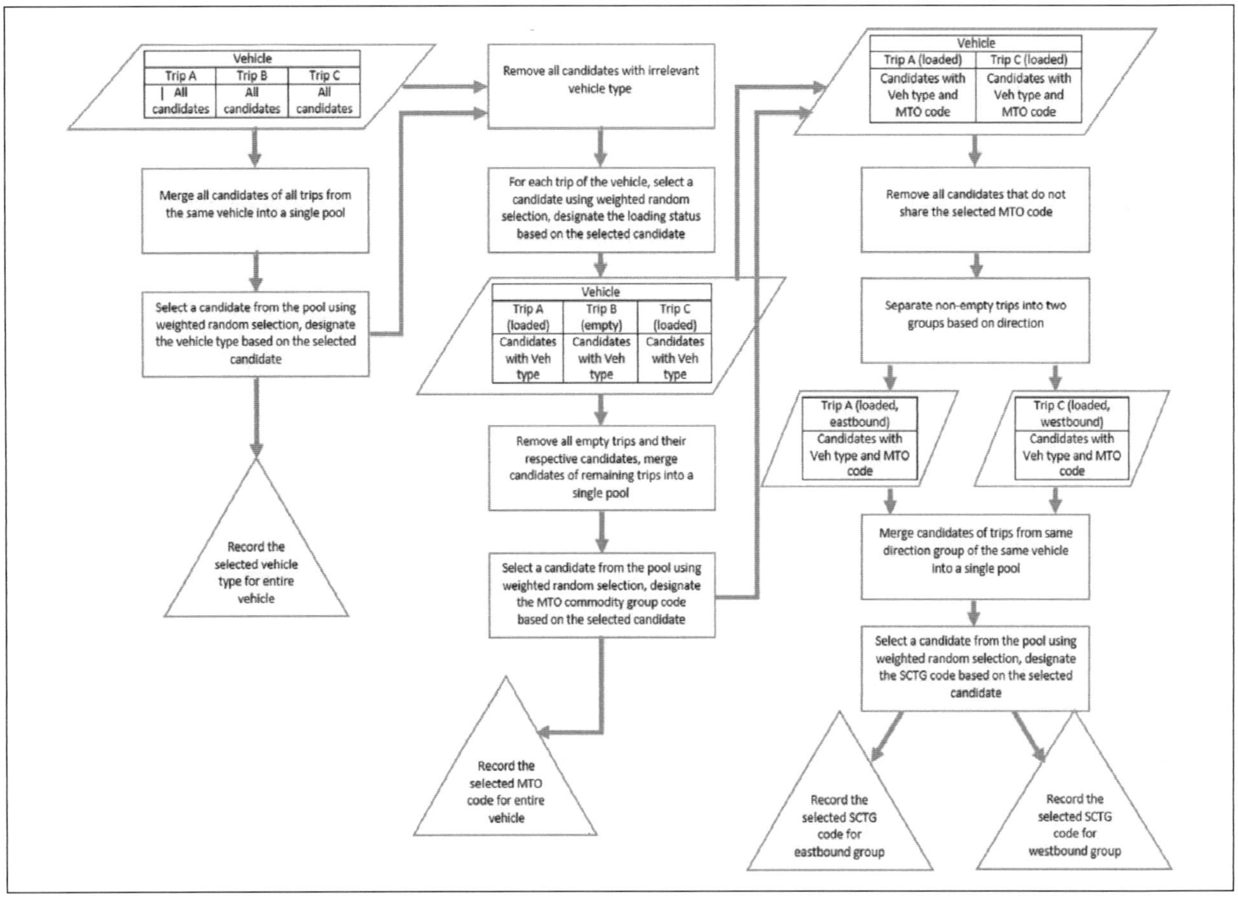

Figure 2. Flowchart illustration of incremental attribute imputation.

Finally, a pool of CVS records is created for each direction, a CVS candidate is selected for each GPS trip from the appropriate directional pool, and other attributes of values of cargo weight, capacity status, detailed SCTG commodity code and trip type are imputed at the individual trip level.

Results

Data Preprocessing and Determining CVS Candidates

The GPS data cleaning and trip generation process resulted in a total of 35,260 trips in the GPS data, which clearly captures a greater proportion of trips compared with the 10,172 CVS records. For the generated GPS trips:

- 1,837 (5.2%) trips found a matching CVS record with trip origin and trip destination within 200m proximity of GPS trip origin and trip destination;
- Allowing other stop combinations (i.e., CVS previous stop and next stop) in addition to CVS trip origin and destination, 15.3% of GPS trips found a CVS record with matched origin and destination;
- Lastly, allowing either origin or destination GPS trip end to be matched to a CVS record within 200m proximity, 85.5% of the GPS trips had their origins matched to some CVS records at the most disaggregate spatial scale, and 90.5% of the GPS trips had their destinations matched to some CVS records at that spatial scale.

Figure 3 depicts the increase of number of candidates with respect to spatial scale. The additional CVS records that become qualified as candidates through spatial scale aggregation are considered "new candidates" and added to the pool at each spatial scale.

Evaluation of Imputation Results

Two evaluation approaches are used to confirm the accuracy of imputed data: internal and external. Internal evaluation relies on data that are already used in the imputation process to confirm the imputed results. Internal evaluation 1 compares the attribute distribution between imputed GPS trips and CVS data. Internal evaluation 2 is a holdback evaluation method. External evaluation uses a third data source that is

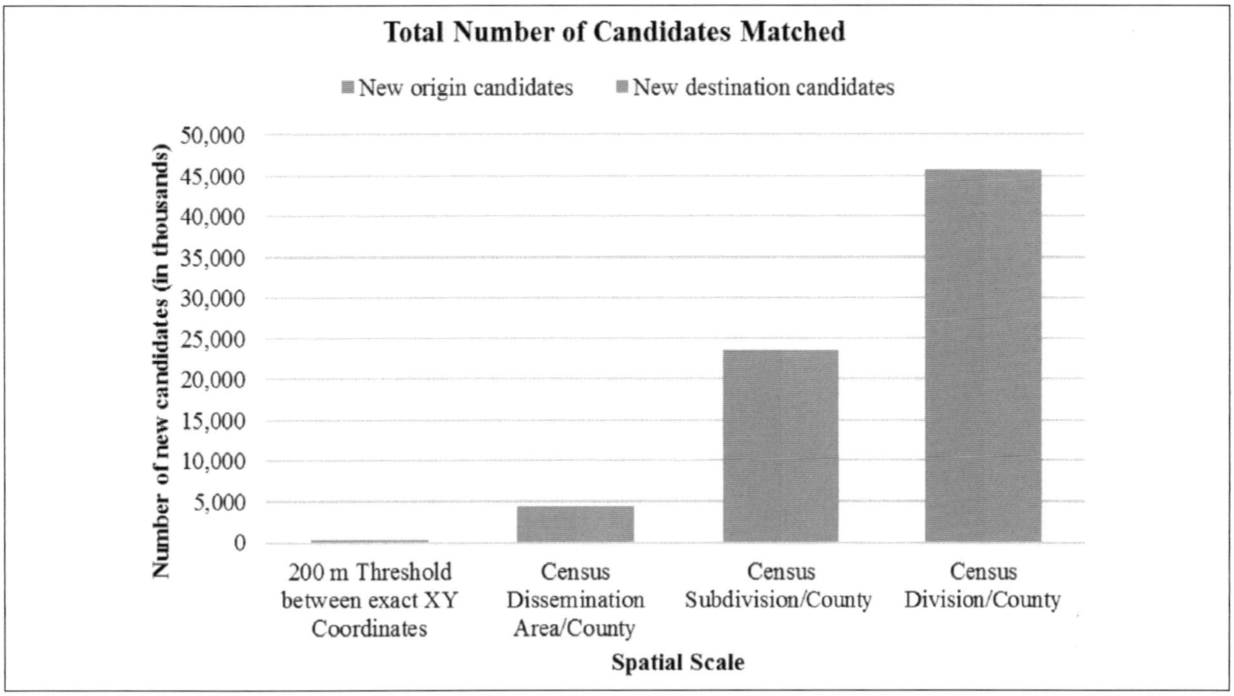

Figure 3. Number of candidates for different spatial scales.

independent of donor data and receptor data to evaluate the performance of imputed data. The third data source that is available for this study is the highway traffic count of vehicles passing the Trafalgar DCS over the week when the CVS was collected at that DCS, which distinguishes between single-unit and multi-unit trucks. For our external evaluation, truck configuration types 4, 6, and 95 are assumed to be single-unit vehicles and the rest are multi-unit vehicles.

Evaluations of the imputation results for each of the three penalty score variations (defined in Table 1) are presented in the following sections.

Evaluation of Imputation Results for Vehicle Type Attributes. Table 3 summarizes the results of the internal and external evaluation for vehicle type attributes. Internal evaluation 1 uses the mean absolute error (MAE), mean relative error (MRE), and root mean squared error (RMSE) to measure the performance of each variation. The preferred variation is the model with the lowest error values. Table 3 shows that Variation 2 is the best performing model for assigning the vehicle type attribute based on truck configuration only. Considering the combination of truck configuration and body style, Variation 3 performs the best and Variation 2 is ranked second.

Internal evaluation 2 assesses performance using a randomly selected holdback sample of 5% of CVS records. Attributes are removed from the CVS records in the holdback sample. These records are run through the imputation process as if they were GPS trips. The imputed attribute values are then compared with the true attribute values. Table 3 shows that all model variations have over 50% accuracy for truck configuration and body style combination. Variation 2 has the highest correct match ratio for truck configuration and body style combination (75.6%) and has approximately the same performance as Variation 1 for match of truck configuration only.

External evaluation compares imputed results with the vehicle type distribution obtained from the highway traffic counts. The largest error is for eastbound traffic, for which the imputed distribution differs from observed by 3.7%. For westbound traffic, its imputed distribution only differs from observed by 1.6%.

The performance of each variation is shown in Figure 4, which shows that Variation 2 has the most consistent performance across all truck configurations. It is more closely aligned to the CVS data in more truck configurations than the other two variations. The only outlier is for truck configuration and body style combination of 1-2 (tractor plus one refrigerated trailer), which differs from expected value by a larger margin. Based on the error values, holdback results, and visual assessment from the graphs, it is concluded that Variation 2 has the best performance amongst the three variations for vehicle type imputation.

Evaluation of Imputation Results for Commodity Type and Trip Type Attributes. Imputed data at the commodity and trip level are also evaluated using the same method described above. General findings are summarized as follows.

Table 3. Internal and External Evaluation Results of Vehicle Type Attributes

Internal evaluation 1: Error of attribute distribution between CVS and imputed data

Model variation	Truck configuration only distribution				Truck configuration and body style distribution			
	MAE	MRE	RMSE	Ranking	MAE	MRE	RMSE	Ranking
Variation 1	2.29×10^{-2}	87%	3.28×10^{-2}	3	4.16×10^{-3}	69%	9.27×10^{-3}	3
Variation 2	0.81×10^{-2}	39%	1.12×10^{-2}	1	4.02×10^{-3}	70%	8.98×10^{-3}	2
Variation 3	1.19×10^{-2}	39%	1.85×10^{-2}	2	3.37×10^{-3}	64%	7.13×10^{-3}	1

Internal evaluation 2: Holdback evaluation of truck configuration and body style

Model variation	% of holdback sample	% of truck configuration matched	% of both truck configuration and body style matched
Variation 1	5%	87.0%	67.8%
Variation 2	5%	86.8%	75.6%
Variation 3	5%	75.8%	55.4%

External evaluation vehicle types with highway traffic count data

	Eastbound		Westbound	
	Single-unit vehicle	Multi-unit vehicle	Single-unit vehicle	Multi-unit vehicle
Traffic count data	20.2%	79.8%	22.4%	77.6%
Variation 1	29.4%	70.6%	28.8%	71.2%
Variation 2	23.9%	76.1%	24.0%	76.0%
Variation 3	24.9%	75.1%	25.2%	74.8%

- For commodity attribute imputation:
  - Error values are generally greater than those of vehicle type assignment. This outcome is expected because commodity type is imputed at a lower level of trip aggregation.
  - External evaluation is not conducted because of the lack of an external data source for comparison of the commodity type imputed results.
  - Variation 1 has the best performance amongst the three variations for commodity type imputation.
- For individual trip attribute imputation [including empty versus loaded, commodity weight, detailed commodity class and trip type (line haul versus peddle run)]:
  - There is generally a small group of CVS records remaining that satisfy all attribute values assigned in the previous levels.
  - Evaluation of the distribution of these attributes is done at the aggregate level only because of a lack of observed data for each individual trip.
  - Although the outputs from different imputation model variations are similar, there is a noticeable difference between CVS data and imputed data.
  - For each of the three model variations, less than 40% of the sample is assigned the same CVS records that the holdback samples derive from.
- Overall, Variation 2 generally performs best and Variation 3 almost always performs worst. In other words, power growth of the penalty score gives better results than exponential growth of the penalty score.

Conclusion

The objective of this study is to develop an approach to enrich the large sample size of GPS fleet tracking data with the detailed trip information of CVS data. Several data fusion methods are reviewed and assessed for their applicability to the study. The study uses a hybrid of simple hot-deck imputation method and k-nearest neighbor (k-NN) imputation method from the implicit data fusion approach to assign the attribute values of interest from CVS data to corresponding GPS trips. Three variations of the significance factor formula are developed and evaluated through a set of internal and external evaluation methods to determine the best performing model. The final outputs of the best performing model have a satisfactory level of resemblance to observed attribute values from CVS data at the aggregate level.

Key findings from this research are:

- It is difficult to fill gaps in GPS data with CVS data if a high degree of spatial precision is required when joining the two data sets. However, reducing the degree of spatial precision potentially introduces errors into the imputed data. A balance needs to be found between the

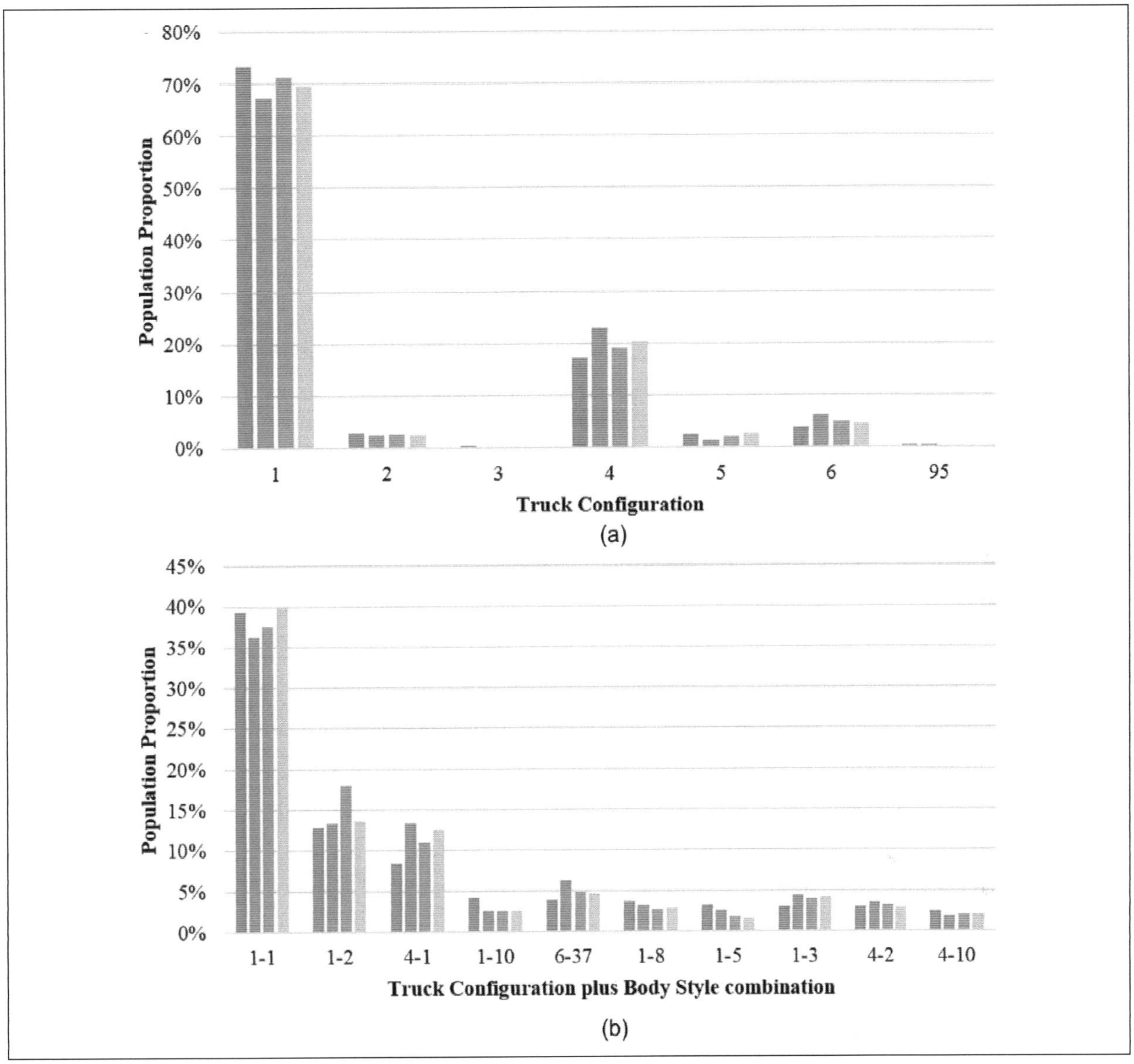

Figure 4. Distribution of (*a*) truck configuration, and (*b*) top 10 truck configuration and body style combinations for imputed GPS data.

two extremes. The balance point varies depending on the type of attribute that is to be imputed.

- The quality of imputation results depends on the number of common variables available between the donor and receptor data. Increasing common variables allows refinement of matched candidates through better definition of "closeness" between the donor and receptor. Data can be imputed at more disaggregated level with more common variables available between GPS data and CVS data.
- The distributions of vehicle type and commodity type of imputed GPS data reasonably reflect those of the CVS data and observed distributions from highway traffic counts.

The imputed data could give support in the future conduct of the commercial vehicle survey. It can identify activity data gaps within the current CVS data, particularly the lack of coverage of urban freight tours. Analysis of the GPS data fusion process could also provide insights to determine locations where new DCS of roadside intercept survey would be most effective, and single out existing DCS that provide limited value for discontinuation or resource reduction.

The study used only a small portion of all GPS fleet tracking data in Ontario. A reasonable next step would be to expand the receptor data to all GPS data within Ontario. There is also the potential to apply the model in other jurisdictions where suitable GPS data and commercial vehicle roadside intercept data are available. Potential also exists to modify the data fusion methodology to enrich GPS data with attributes from other data sources, such as establishment data.

Acknowledgments

The authors would like to thank Ministry of Transportation Ontario (MTO) for the data and financial support. Specifically, they would like to express their gratitude to Rob Tardif of the Systems Analysis and Forecasting Office.

References

1. Van Der Puttan, P., J. N. Kok, and A. Gupta. *Data Fusion through Statistical Matching*. MIT Sloan Working Paper No. 4342-02; Eller College Working Paper No. 1031-05, 2002. http://dx.doi.org/10.2139/ssrn.297501.
2. Bachmann, C., B. Abdulhai, M. Roorda, and B. Moshiri. A Comparative Assessment of Multi-Sensor Data Fusion Techniques for Freeway Traffic Speed Estimation Using Microsimulation Modeling. *Transportation Research Part C: Emerging Technologies*, Vol. 26, 2013, pp. 33–48. https://doi.org/10.1016/j.trc.2012.07.003.
3. Ministry of Transportation Ontario. *MTO - University of Toronto GPS Probe Data Research Initiative*, 2015.
4. D'Orazio, M., M. Di Zio, and M. Scanu. *Statistical Matching: Theory and Practice*. John Wiley & Sons, New York, 2006.
5. Bayart, C., P. Bonnel, and C. Morency. Survey Mode Integration and Data Fusion: Methods and Challenges. *Proc., 8th International Conference on Survey Methods in Transport*, Annecy, France, 2008.
6. Scheuren, F., and W. E. Winkler. Regression Analysis of Data Files that are Computer Matched. *Survey Methodology*, Vol. 19, No. 1993, pp. 39–58.
7. D'Ambrosio, A., M. Aria, and R. Siciliano. Robust Tree-Based Incremental Imputation Method for Data Fusion. *Proc., 7th International Symposium on Intelligent Data Analysis*. Ljubljana, Slovenia, 2007, pp. 174–183.
8. Rudra, M. *Application of Imputation Methods in the Analysis of Freight Trip Generation in the Greater Toronto and Hamilton Area*. MASc thesis. University of Toronto, Ontario, Canada, 2014.
9. Aluja-Banet, T., J. Daunis-i-Estadella, and D. Pellicer. GRAFT, A Complete System for Data Fusion. *Computational Statistics & Data Analysis*, Vol. 52, No. 2, 2007, pp. 635–649. https://doi.org/10.1016/j.csda.2006.11.029.
10. Miller, E., K. N. Habib, M. Lee-Gosselin, C. Morency, M. Roorda, and A. Shalaby. *Changing Practices in Data Collection on the Movement of People*. Final Report for Transportation Association of Canada (Project report, TAC-ATC), 2014.
11. McCabe, S., H. Kwan, and M. Roorda. Comparing GPS and Non-GPS Survey Methods for Collecting Urban Goods and Service Movements. *International Journal of Transport Economics*, Vol. 40, No. 2, 2013, pp. 183–206.
12. Sharman, B., M. Roorda, and K. N. Habib. Comparison of Parametric and Non-Parametric Hazard Models of Stop Durations on Urban Commercial Vehicle Tours. *Transportation Research Record: Journal of the Transportation Research Board*, 2012. 2269: 117–126.
13. Zhu, S. *Integration of Commercial Vehicle GPS and Roadside Intercept Survey Data*. MASc thesis. University of Toronto, Ontario, Canada, 2016.
14. 2017 Commodity Flow Survey Standard Classification of Transported Goods (SCTG). *SCTG Commodity Codes*. U.S. Department of Transportation, Washington, D.C., 2016. https://www.census.gov/econ/cfs/2017/CFS-1200_17.pdf.

The Standing Committee on Freight Transportation Data (ABJ90) peer-reviewed this paper (18-00872).

Article

TRR
JOURNAL OF THE TRANSPORTATION RESEARCH BOARD

Transportation Research Record
2018, Vol. 2672(44) 21–31

DOI: 10.1177/0361198118796024
journals.sagepub.com/home/trr
SAGE

Assessment of Resource Allocation and Tradeoff Analysis Approaches in Transportation Asset Management

James Bryce[1], Gonzalo Rada[1], Samuel Van Hecke[2], and Joseph Zissman[3]

Abstract

Transportation asset management (TAM) practices continue to grow and to develop as transportation agencies seek to make more objective and defensible decisions, as well as responding to recent legislation. One primary goal of TAM is to provide a structure in which decisions on how to distribute resources across many disparate assets can be made using a systematic process. Resource allocation is analogous to multi-objective optimization, and thus presents the complication that many potential optimal solutions (i.e., a Pareto set) can be found. To solve the multi-objective resource allocation problem, many approaches have been recommended, such as the use of utility theory and weighting functions to express preferences that result in a single solution being selected. This paper discusses those recommendations, as well as describing the current state of the resource allocation process in an effort to identify gaps in practice and, in turn, to provide recommendations for addressing those gaps. First, the process of cross-asset resource allocation is deconstructed to highlight the key steps. Then, current practices identified from the literature, as well as from interviews with five State agencies are discussed. Finally, a set of recommendations for improving resource allocation and the supporting processes within resource allocation are presented.

Transportation Asset Management (TAM) is a process that enables departments of transportation (DOTs) to use engineering and business practices to make more informed and transparent investment decisions (*1*). An important aspect of successful TAM implementation within a DOT is a clear relationship between the performance of assets classes and allocation of resources among asset classes. Maturing this relationship requires asset performance to be clearly defined and tied to quantitative measures in such a way that enables DOTs to embrace data driven decision making, to track progress toward goals and to develop performance targets. However, developing the relationship between performance and resource allocation requires the performance of an asset be clearly defined, and for measures to be developed to support the monitoring of this performance. Additionally, once the relationship between the asset performance and resource allocation is defined for each asset, DOTs must then develop a means to compare the consequences of investing in multiple, disparate types of assets.

Two significant components of TAM are cross-asset resource allocation and tradeoff analysis, which are processes that agencies use to assess the effect of varying funding levels over several groups of assets and allocate resources to each asset. Commensurately, the objective of this paper is to describe the current state of TAM resource allocation and tradeoff analysis by examining many sources of literature, as well as practices among State DOTs that are determined based on interviews. The basic steps to implementing robust resource allocation practices are presented, and a path forward to addressing challenges within the current state of the practice are detailed. The content of this paper is based on broader research conducted by the authors for the US Federal Highway Administration.

Defining Transportation Asset Management

TAM is a broad concept, and as a result, has many different definitions. In 2002, the *Transportation Asset*

[1]Amec Foster Wheeler, Beltsville, MD
[2]Cambridge Systematics, Chicago, IL
[3]Cambridge Systematics, Medford, MA

Corresponding Author:
Address correspondence to James Bryce: James.Bryce@amecfw.com

Management Guide defined TAM as "a strategic approach to managing physical transportation infrastructure assets" (*2*). In 2006, the National Cooperative Highway Research Program (NCHRP) Report 551 defined TAM as "a set of principles, concepts, and techniques for making more effective allocation and use of resources to address preservation, operation, and improvement of transportation infrastructure" (*3*). In 2013, the American Association of State Highway Transportation Officials (AASHTO) *Transportation Asset Management Guide – A Focus on Implementation* defined TAM as "a strategic and systematic process of operating, maintaining, upgrading, and expanding physical assets effectively throughout their lifecycle" (*4*). In 2015, MAP-21 defined TAM as "a strategic and systematic process of operating, maintaining, and improving physical assets, with a focus on engineering and economic analysis based on quality information, to identify a structured sequence of maintenance, preservation, repair, rehabilitation, and replacement actions that will achieve and sustain a desired state of good repair over the lifecycle of the assets at minimum practicable cost. (23 U.S.C. 101(a)(2), MAP-21 1103)" (*5*).

Although each definition of TAM is unique, there are common elements that permeate the group of definitions, including "*strategic and systematic*," "*throughout their lifecycle*," "*quality information*," and "*well-defined objectives*." In essence, TAM allows DOTs to operate rationally and comprehensively, by "*starting with clear strategies for what infrastructure conditions they want to provide to the public*" (*6*). TAM is able to promote more effective resources allocation and utilization based on quality information within a state DOT or other regional/local transportation department.

Resource Allocation and Tradeoff Analysis

According to NCHRP Report 806, cross-asset resource allocation is a "framework to optimally distribute the budget across multiple assets such as pavements, bridges, culverts, guardrails, and signals, and to promote the quality of service of the entire facility" (*7*). Many DOTs include impacts on environment, land use, greenhouse gas emissions, recycling practices, and material consumption, for instance, within their asset resource allocation processes (*8*). NCHRP Report 806 details two primary approaches that DOTs can use to link performance-based planning to cross-asset resource allocation decisions: target-based allocation and preference-based allocation (*7*). Target-based allocation is a qualitative approach that occurs within one asset class, and will not be detailed in this paper. One of the most common preference-based allocation tools used by DOTs is multi-objective decision analysis (MODA), which uses a decision science approach, based on preferences that "guide project selection through weighting, scaling, scoring, prioritization, and optimization techniques" (*7*). To compare projects either within or across asset management systems, MODA assigns a representative score that combines DOT weighting and scaling preferences. The scores indicate the relative importance of pursuing a project and can be used to support prioritization efforts.

Another common preference-based allocation tool that is often used with MODA is trade-off analysis, which can be used to further inform the decision-making process by evaluating the impacts of various investment levels and allocation strategies on overall system performance, as well as the extent to which one performance measure can be exchanged for another (*9*). Trade-off analyses inform decision makers of the consequences (i.e., what is gained and what is lost) of moving from one feasible solution to the next. While state DOTs have become proficient in managing bridge and pavement assets independently, findings from NCHRP Report 545 indicate that DOTs struggle with comparing "apples to oranges" trade-offs across asset classes (such as mobility versus preservation and maintenance versus expansion) (*10*).

When defining methods for performing cross-asset resource allocation, it is important to understand the general structure of the resource allocation process. For each of the four phases below, a DOT has several alternative strategies, some are more data-driven than others. To distribute its limited resources, a DOT must perform each of the following steps:

1. Generate project proposals: An allocation approach that prioritizes and funds a poorly conceived or incomplete universe of possible work will not generate an optimal work program. Project proposals are asset, corridor, or region-level maintenance or capital tasks under consideration for the work program. Proposals can be fully scoped, scheduled, and budgeted projects with estimated benefits, or can be conceptual treatments that achieve established performance goals. Examples of how proposals can be generated include:
   a. Relying on the judgment of staff and program management. DOT engineers are typically deeply familiar with the characteristics and deterioration of assets under their purview and have strong intuition about their deterioration and response to treatments. They may also possess institutional knowledge about management and stakeholder response to certain types of proposals.
   b. Incorporating stakeholder input from local governments (e.g., counties, cities, or towns)

and from metropolitan planning organizations (MPOs), who are required by Federal law to produce a transportation improvement plan (TIP) that allocates funding at the regional level. DOTs vary in how they adopt or relate to these plans from adopting them wholesale (thereby allowing most/all prioritization to occur at the local level) to strongly cooperating with MPOs and exercising significant influence over the TIP process.

   c. Using an asset management system that procedurally identifies and prioritizes current and future deficiencies using an inventory and condition database supported by inspections. These systems often reference long-term deterioration curves based on local and/or national datasets to evaluate the impact of different treatments on asset conditions. They can also account for the typical schedule of treatments that a DOT deploys over an asset's service life.

   Agencies can (and in many cases do) use all three of these approaches in tandem.

2. Distribute funding among programs: A DOT may draw from dozens of individual funding sources including, but not limited to Federal Aid, State allocations and matching funds, and user fees (e.g., tolls and transit fares). Many of these sources of funding arrive with designated uses and restrictions. Likewise, programs may have been guaranteed a certain level of funding by legislative action. When there is funding flexibility, distribution strategies include:
   a. Rely on precedent, traditional and expert knowledge to provide each program with an expected amount of funding.
   b. Use funding formulae, algorithms, or other quantitative approaches that are not responsive to performance data (e.g., geographic distribution or a firm percentage for preservation).
   c. Develop a distribution strategy based on performance data, in which funding is applied to programs that will yield the most value.

   NCHRP Report 806 discusses an alternative that eliminates this phase by removing programmatic considerations and applying funding directly to projects. However, the recommended approach in this paper uses a performance-based distribution process. This process is preferred over distributing funds directly to projects because programs can identify additional projects over and above the initial proposals if they can pay for them under the recommended allocation. In other words, if a specific program generated proposed projects with an associated cost, a bottom–up approach can only assign that program a maximum of the associated cost from the proposed projects, while a top–down approach can assign it an amount greater than the associated cost.

3. Prioritize projects for inclusion in the work program: Project selection can occur either within a program after funding has been distributed or across the agency within the full project universe. DOTs must consider in this phase whether a proposal can be realistically achieved within the work program's full multi-year time frame. Outside of that concern, it is also possible to, with adequate data, consider costs, benefits, and whole-life costs of projects.

   Strategies for project prioritization resemble those for proposal generation – judgment of DOT engineers, MPOs and local control, and use of an asset-management system, with a combination of these being the most common.

4. Schedule the work program: Once the full multi-year project portfolio is selected, projects must be scheduled into each fiscal year. In this stage, the DOT must consider funding projections for each individual year, as well as the likelihood that individual projects will be designed, approved, and shovel-ready. Generally speaking, this is an arduous process that is performed manually by informed staff, either within a DOT or in MPOs and local authorities.

Resource allocation problems often require multiple iterations, in which the solution from one evaluation informs subsequent allocations. For example, Dehghani et al. proposed a cross-asset management framework consisting of the following steps that are repeated until the performance evaluation results in desirable outcomes (*11*):

1. Resource allocation. The process begins with an initial allocation scenario based on expert opinions, goals and objectives, and DOT constraints. In this framework, the results and outcomes of each resource allocation scenario are compared with the DOTs goals and objectives, and are used to update the scenario until the optimal resource allocation is reached.
2. Treatment selection. Once the DOT selects a resource allocation scenario, the resources are distributed across its assets. The DOT then selects the best maintenance treatments based on the available budget. Generally, the treatments can be

classified under the main categories of preventive, corrective, restorative, and heavy rehabilitation.

3. Performance prediction. After the DOT selects the treatments, the next step is to predict performance improvements from maintenance treatment applications. Functional performance represents an evaluation of the asset serviceability and superficial conditions (i.e.; roughness, surface friction, etc.) and structural performance identifies asset distresses and structural deficiencies. These two performance types can be combined with other performance features such as a safety or environmental evaluation to develop a combined performance.
4. Overall performance evaluation. The predicted performance measures for each asset are aggregated into combined measures that explain the entire corridor or facilities performance.

The approach recommended by Dehghani et al. represents one method for enhancing the resource allocation step by assessing the effect of various levels of allocation on individual assets or the entire system (*11*). This approach results in allocations that are based on an evaluation of system performance, instead of basing allocations on historical precedent. although some states have two-step methodologies that straddle or combine these categories, approaches to distribute funding among programs generally fall into the following four categories (*7*):

- Legacy driven: The approach to resource allocation for several DOTs is largely legacy driven, meaning that the amount of funding that goes toward addressing a given goal or objective is largely determined by the existing program structure and either the amount or share of funding that has historically gone to each program. Changes may be made at the margins to accommodate inflation or emerging priorities.
- Preserve it first: Several state DOTs have allocation processes driven largely by asset management systems that determine needs for preservation investments. Owing to financial constraints in these states, this tends to leave very little funding left to be allocated for other purposes. Remaining allocations are thus typically a reflection of desired projects rather than goal-level priorities.
- Soft optimization: To varying degrees, some states use approaches that may start from a legacy or fix-it-first approach, but then adjust allocations to better align spending with the relative priorities of the DOT. This is called "soft optimization" because decisions are mostly or partly driven by professional discretion and other nontechnical inputs.
- Performance-based: The states that come closest to performance-based cross-asset resource allocation are those that have heavily integrated performance measurements into their resource allocation decision-making approaches. The main difference between this category and soft optimization is that decisions are more data- and analysis-driven through the use of models and tools that help the state DOT forecast likely outcomes across different goal areas. Ultimately, allocation decisions still require leadership to make professional judgments about the right balance of trade-offs.

Each of the above four categories represent varying degrees of analytical rigor, with a performance-based approach requiring data-driven approaches. Similarly, the approaches to conduct resource allocation can be implemented with varying degrees of analytical rigor. As demonstrated in Figure 1, cross-asset resource allocation approaches can generally be formulated using one of three approaches (*7*):

- Traditional approach: The decision-maker decides how to split x number of dollars among n number of assets or programs based on past experience and input from subject matter experts (e.g., program managers for the specific assets). This approach relies on assumptions and precedent to distribute funding to siloed programs (i.e., programs viewed in isolation of each other).
- Top–down approach: The decision-maker decides how to split x number of dollars among n number of assets or programs based on an understanding of how each asset or program area will respond to a given funding level. The response (performance) of an asset or program area to a given level of funding can be developed in many ways, as is described in later sections of this paper. The responses are used to generate response curves (performance measure as a function of funding level), which provide immediate feedback about the implications of levels of funding.
- Bottom–up approach: The decision-maker decides how to split x number of dollars among n number of assets based on selecting a portfolio of projects across all assets which maximizes the overall utility (measure of preference) of the decision-maker. This approach begins with specific projects, and then groups projects across various assets so that the overall sum of the utilities is maximized (subject to constraints defined by the analyst). Unlike the top–down approach, there is no direct feedback about the implications of the level of funding

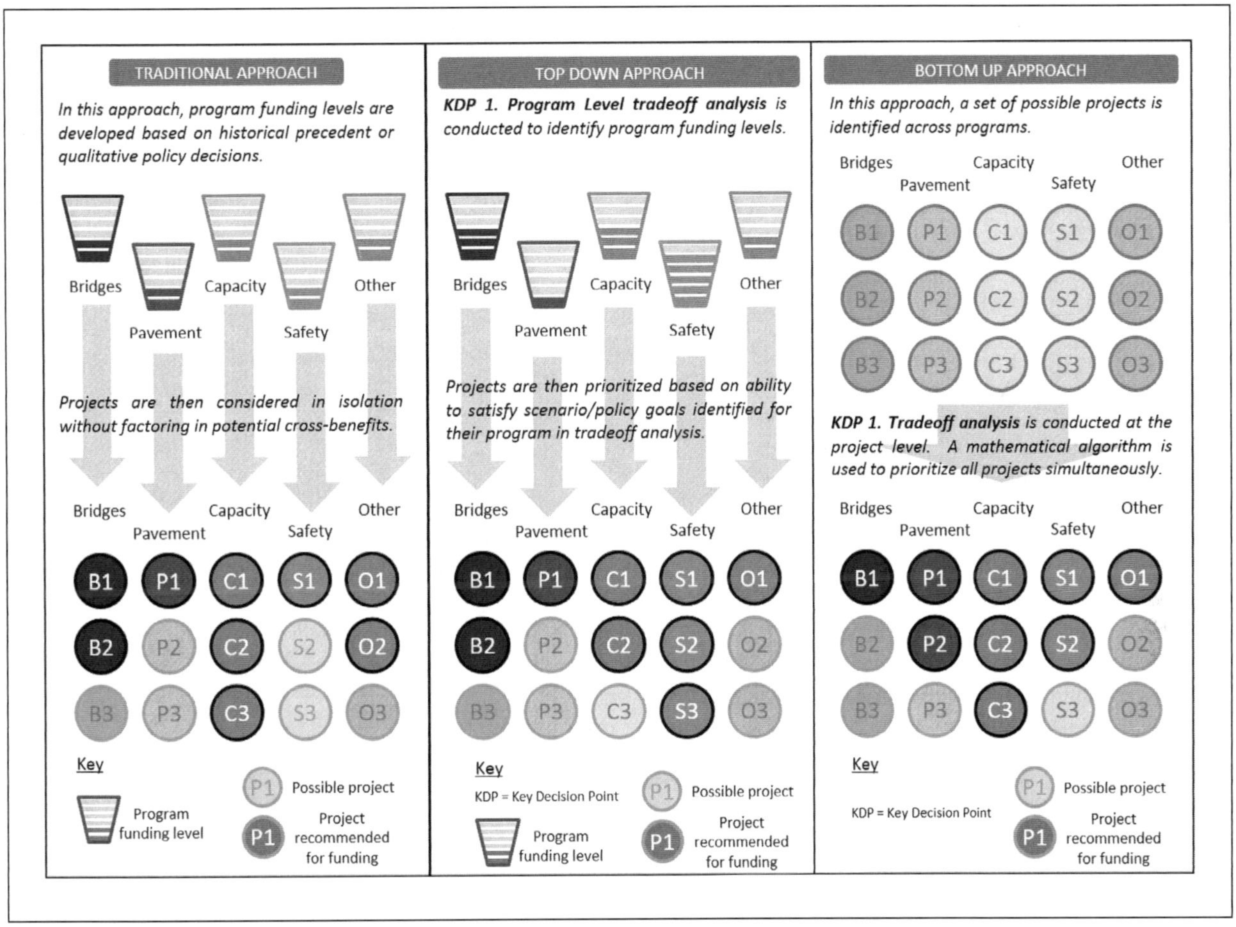

Figure 1. Typical approaches to cross-asset resource allocation.

applied to an asset class. Instead, it must be assumed that the benefits of selecting particular projects are scaled and weighted in such a manner that the incremental benefits of selecting additional projects diminishes (or stays constant in one specific case) as the number of selected projects within an asset increase.

Sample Practices

Pavement and asset management personnel from five State DOTs were interviewed as part of the research documented in this paper to discuss the characteristics they view as important for TAM, as well as to gain an understanding of their TAM practices. The DOTs that were interviewed were: Arizona DOT, Indiana DOT, Kansas DOT, Maine DOT and Vermont DOT. Below are a sample of practices based on information provided via interviews with personnel from the Maine and Indiana DOTs, which specifically discussed having formalized approaches to resource allocation among different assets.

Maine. Project selection is generally the most critical opportunity for cross-asset resource allocation decisions. Maine DOT varies project selection methodologies by asset type and mode – prioritization is performed by the Highway Committee, Bridge Committee, and Multimodal Committee, all comprised of technical staff (*12*). The committees use a mixture of quantitative and qualitative analysis to score and select projects.

In principle, Maine DOT prioritizes corridors, not projects. The Highway Corridor Priority (HCP) score for roadway segments is based on functional classification (State and Federal), truck traffic, relative amount of traffic, and level of dependent business activity. HCP is a six-point scale, with interstates rated HCP 1 and local roads rated HCP 6. The HCP scale is combined with the Customer Service Level (CSL) to grade corridors on an A–F scale. CSL is based on metrics for safety (e.g., crash history, paved roadway width, pavement rutting, and bridge reliability), condition (e.g., pavement condition, roadway strength, bridge condition, and ride quality), and service (e.g., posted road, posted bridge, and congestion) (*3*). The Highway and Bridge Committees consider the A–F grade and CSL independently when selecting candidate projects. In addition, they consider qualitative questions, including the following:

- What is the customer benefit that will arise from the candidate project? Are there other transportation alternatives if this candidate project does not proceed?
- Does the candidate project address a demonstrated safety issue?
- Does the candidate project involve a public-private or public-public partnership that will stretch scarce transportation resources?
- Does the candidate project relate to economic activity and job growth?
- Does the candidate project fill a gap in the existing highway corridor or other asset?
- Is the candidate project able to proceed to construction in the next three years, considering environmental documentation and permissions?
- Is there public support for the candidate project?
- Does the potential project preserve an element of the existing transportation system, such as replacing a failing bridge or protecting an existing transportation investment?

Cross-asset allocation is performed at the front-end – Maine DOT reports that funds are divided between modes and programs based primarily on legacy commitments and funding source restrictions. Once funds have been allocated to the three committees, they determine the Work Plan largely independently.

Indiana. Indiana DOT (INDOT) has a "capital program management" process based around four "capital program management groups" (or "asset management teams") that have been given latitude to develop business rules and scoring factors for different types of projects. The groups include:

- Traffic mobility: This team prioritizes projects that serve established state goals for travel congestion, accessibility, and connectivity. This is composed of nine unique work categories, including the addition of travel lanes, new road construction, intersection improvement, and interchange modification. The intersection projects are considered jointly with the traffic safety team.
- Roadway/pavement: This team prioritizes projects that preserve or modernize road geometrics and pavement, with special attention paid to minimizing life cycle costs (which INDOT refers to as "sustainability"). Projects include pavement preservation, pavement resurfacing, and roadway/pavement modernization.
- Bridge: This team prioritizes projects that maintain, enhance, and modernize existing bridges and small structures. Projects can range from pipe lining to deck overlay and bridge replacement.
- Traffic safety: This team prioritizes projects that improve safety on INDOT roads. The scoring system favors the most cost-effective investments with regards to severe injury and fatal crashes. Work types include the installation of new or improved traffic control devices and the reconstruction of intersections and road segments, among others. As stated earlier, some projects are considered jointly with the traffic mobility team.

INDOT monitors its transportation systems (i.e., the sets of assets that correspond to the asset management teams) continuously for under-performing assets and services. It uses these performance reports to identify assets in need of investment over the upcoming five-year planning horizon. Candidate projects are primarily developed by districts, and are submitted during an annual call for projects.

Candidates are scored on a 0–100 scale by their respective committees. The specific criteria and formulae (i.e., business rules) are determined by the committee itself. Typically, scores are based on between five and ten factors that capture both the need and the expected impact of the corrective action. Scoring is also designed to be consistent with long-term performance goals, established by the long range transportation plan (LRTP) and by other INDOT policy documents. The asset teams divide candidate projects into A, B, C, and D groups, which correspond to the quartiles of the 0–100 scores. They then review each project and adjust the ratings as necessary, removing some from consideration. Once revised and confirmed, the lists are forwarded to the program management group (PMG) for cross-asset resource allocation. The PMG distributes funding with an awareness of trade-offs between programs and the purchasing power of investment in different programs. The approach by Indiana is an example of a bottom–up approach.

Common Approaches

DOTs face a particularly difficult task in balancing the diverse set of needs involved in maintaining a complex transportation system. The evolution of TAM and performance-based planning and programming has resulted in a set of analytical approaches and tools to assist DOTs with resource allocation. These efforts have further supported the foundation for frameworks that link system performance to resources. Cross-asset resource allocation offers an opportunity for decision-makers to consider the consequences of investment options during a period when needs exceed available resources. Some common tradeoffs found in literature and from the interviews include those between:

- Types of assets (e.g., bridges and pavement)
- Expanding the system, modernizing it, and maintaining it in a state-of-good-repair
- Sub-systems (e.g., interstates vs. national highway system vs. non-interstates)
- Geographic regions
- Program areas (safety vs. mobility)
- Modes (e.g., highways vs. transit)

In a fiscally constrained environment, deciding to fund one priority directly prohibits the funding of others. These decisions, which are made on a regular basis by transportation professionals and elected officials, have significant implications on system performance. An objective, data-driven tradeoff analysis provides a structured approach for evaluating and communicating the implications of various funding decisions.

In addition to improving funding decisions, tradeoff analysis can enable DOTs to set meaningful performance targets. Interest in performance targets, as an organizing principle for transportation planning and programming, has been solidified by guidance provided in the MAP-21 Act and the FAST Act (*5*, *13*). In general, implementing tradeoff analysis within a DOT requires:

- Clear goals and performance objectives
- Performance measures
- Data and analytical tools to evaluate the program and project-level decisions with regard to system performance
- A decision analysis process that takes advantage of the analysis

Most DOTs incorporate TAM and/or performance-based principles in some form. However, resource allocation has traditionally been conducted within a single asset type or program area in isolation from other programs. This approach prevents a detailed understanding of the implication of increasing funding for one program against another (e.g., pavement preservation vs. highway expansion vs. bridge preservation). It also can hinder the ability of DOTs to evaluate and communicate the comprehensive benefits of specific projects. For example, capacity expansion projects can affect safety performance as well as future maintenance needs. A pavement preservation project in a corridor can present a low-cost opportunity to rehabilitate bridges. A holistic view of resource allocation further supports getting the most from each dollar spent. The success of any resource allocation process is only possible if the widest range of consequences can be communicated effectively.

Assessment of Approaches and Practices

Each approach and practice detailed in this paper has many advantages and drawbacks, which are discussed in this section. The principal advantage of the traditional approach is predictability; if subject matter experts have been conditioned to expect a certain amount of funding, they may be generating projects and managing expectations based on those assumptions. A sudden reduction or increase in funding could both be equally challenging, as a significant reduction would require considerable change to the project scheduling phase, while a windfall would require the rapid development of additional projects. Predictability can also be expressed simplistically when considering performance; if the system is working within a specific program, it can be difficult to justify changing it. The primary risks associated with the traditional approach is that there is no certainty that the decisions being made represent the best possible decisions; performance and cost-effectiveness and therefore could be optimized by pursuing a more data-driven strategy. It also places a heavy burden on program managers – not only is their project selection process nearly fully-responsible for meeting Federal TAM and performance standards – but qualitative distribution and established policies may make it difficult to achieve them. Engineering judgment can include institutional knowledge that must be personally conveyed or passed during succession: a program where a single person has always prioritized projects "their own way" is not sustainable in the long term.

The bottom–up approach provides consistent allocation across programs and assets, is objective and defensible, and ideally represents TAM principles. The most prominent disadvantage of the approach (as noted in NCHRP Report 806) is that significant technical challenges exist to implementing it. These include the difficulty of developing a robust and universal algorithm, as well as limitations to data collection and quality (which can be overcome in time as the state of practice improves). Beyond the technical challenges, however, an idealized bottom–up approach does not respect statutory constraints on funding and programs or efficiencies that can be achieved by performing certain projects together. In addition, it includes no program-level analysis and, as such, provides no guidance for proposal generation. A DOT that adopts bottom–up prioritization to reorder projects, but continues to develop the same project universe as before will not innovate its priorities. Much like the traditional approach, this relies on subject matter experts to provide quality inputs.

Critical to the bottom–up approach is an understanding of the marginal rate of substitution, or the rate at which the decision-maker would be willing to trade off between the various assets to maintain a specific value of utility. For a network of many assets and many projects, there is likely to be many portfolios of projects that maximize (or come close to maximizing) utility. In economics and decision sciences, the marginal rate of substitution is used to create an indifference curve, and the optimal solution to the resource allocation problem is tangent to the indifference curve. The concept of the indifference curve highlights the importance of forecasting the quantitative consequences of tradeoff decisions, which inform how the marginal rate of substitution changes over many factors, and ultimately inform the prescribed solution.

Utility theory is generally employed in tradeoff analysis, and is recommended for use with the bottom–up approach in NCHRP Report 806. However, many reports that employ utility theory, including NCHRP Report 806, do not specifically discuss the implications of mutual utility independence and, subsequently, additive independence. This assumption states that the range of utilities assigned to one criteria are independent of the values of the other criteria (*9*). As an example, this assumption states that a decision-maker values a certain improvement in pavements (e.g., an improvement from 3–4% poor) the same regardless of the condition of the other assets. In this assumption, the decision-maker does not change the relative utility gained from improving one asset based on the condition of other assets.

It is important to recognize that mutual utility independence is not likely for transportation investments, as different measures of asset condition are often correlated (i.e., as superstructure condition declines and full replacement of the bridge becomes likely, less utility would be placed on keeping the deck in perfect repair). Keeney and Raiffa discuss the implications of the independence axioms, and note the importance of accounting for interdependencies when applying utility theory to multi-objective decision-making (*9*). If interdependence is not accounted for, the optimal solution found by maximizing utility based on selecting specific projects may not represent the 'best' solution based on DOT preferences.

The ability to maintain existing project selection methods when they function well (on a program-by-program basis) is a key advantage of the top–down approach. This approach allows DOTs the flexibility to maintain what works while introducing objective analysis – preference-based, monetized, or linear – to programs where it does not yet exist. Another advantage is that proposal generation has a feedback loop with allocation; if not enough proposals exist to exhaust the optimal funding level, more can be generated. On the other hand, proposals and expectations can be tempered to match realistic levels of funding, supported by evidence that those allocations are ideal for the DOT. A disadvantage of the top–down approach is that considering the co-benefits of projects (i.e., the added benefits to projects in other assets gained from implementing a project in a given asset) is difficult if strict quantitative project selection is employed for all programs. It may be a challenge to convince program managers presented with a limited amount of funding to respect measures for other programs above their own. It must also be noted that the top–down approach can misrepresent qualitative approaches to project selection as part of a performance-driven process if the response curve is not developed in a fairly robust manner.

Based on interviews with five state DOTs and a review of the literature on tradeoff analysis, the following were identified as implementation barriers or existing limitations to tradeoff analysis practice. Effective tradeoff analysis can be adversely impacted by:

- An overly technical process: A process dependent on jargon, complex mathematical theory, or black-box computation may not be transparent to stakeholders and program staff. If subject matter experts cannot understand how a system works, their expertise cannot be used to improve or refine it.
- Lack of a feedback loop between project prioritization and program distribution: This gap appears most prominently when one considers the cost of delaying projects – performance may not be available at the same price over time. Similarly, a program provided with a larger amount than anticipated may not be prepared to spend it efficiently.
- Too much or too little centralization of decision making: Failure to communicate and implement high-level strategic goals throughout a DOT can lead to executives and stakeholders expecting performance that may be impossible given constraints. Similarly, conflicting sets of priorities – among siloed programs or regional decision-makers – can muddy the waters and challenge the ability of DOTs to meet system-wide performance goals.
- Unclear relative priorities between different goals and objectives: A mature and well-founded set of performance measures cannot support cross-asset resource allocation unless the agency knows which ones it values most.
- Lack of co-benefit analysis: Projects tied directly to one type of asset (or funded through one program) may benefit a wide variety of assets. Siloed project selection may not consider these co-benefits, and prioritization from a single pot – as in a bottom–up approach – may miss them as well if

projects are not scored on an appropriate variety of measures. In addition, many DOT programs address multiple types of assets across a corridor at once, a strategy that can realize cost-efficient benefits.

Discussion and Recommendations

This paper detailed approaches to TAM resource allocation problems considering tradeoffs between various assets. Cross-asset resource allocation offers an opportunity for decision-makers to perform tradeoff analysis while evaluating investment options during a period when needs exceed available resources. Much work has been performed in support of developing methodologies for cross-asset resource allocation, but many challenges still exist that can adversely affect the successful implementation of TAM. As the overall state-of-practice for asset management and performance management improves over time, improvements to the supporting processes must also occur. This includes improvements that move the state-of-practice toward making more defensible decisions on resource allocation. The development of a next generation of asset management methodologies supports this improvement in the state-of-practice.

Based on the assessment of current approaches to TAM resource allocation, it is recommended that, more data-driven approaches should be adopted for resource allocation. However, it is important to maintain a balance between data-driven resource allocation and expert judgment. Institutional knowledge is a significant resource that DOTs maintain, and this knowledge should be reflected in the resource allocation approaches. In other words, a balance should occur so that the adopted techniques are not overly reliant on 'black box' style analytical techniques, but instead the adopted techniques should maintain analytical flexibility that includes methods for capturing the institutional knowledge within an agency.

In an effort to capture the strengths of each approach to cross-assets, a two-step approach was developed and recommended within the research described in this paper. The recommended two-step approach to tradeoff analysis takes into account the observed gaps in sample DOT approaches, as well as the limitations of the three theoretical approaches described in the prior section. The recommended approach also respects many processes that a DOT currently has in place, and provides a framework that can be used to systematically use and improve each of those processes. For example, risk is a critical part of TAM, and many DOTs have developed plans for managing risk (e.g., the impact of extreme weather on assets and system performance). The approach addresses each of the four decision points of resource allocation and is designed to reflect the desire of DOTs to retain the elements of their legacy approach that function well.

The processes within the recommended two-step approach are shown in Figure 2. As previously discussed, the recommended approach draws from the strengths of the three approaches (traditional, top–down and bottom–up) to inform tradeoff analysis. In addition, the two-step approach can facilitate the inclusion of many enhancements (e.g., cross-benefit analysis) that will be discussed later in this section.

The recommended approach primarily consists of the following processes:

- Generate project proposals: An allocation approach that prioritizes and funds a poorly conceived or incomplete universe of possible work

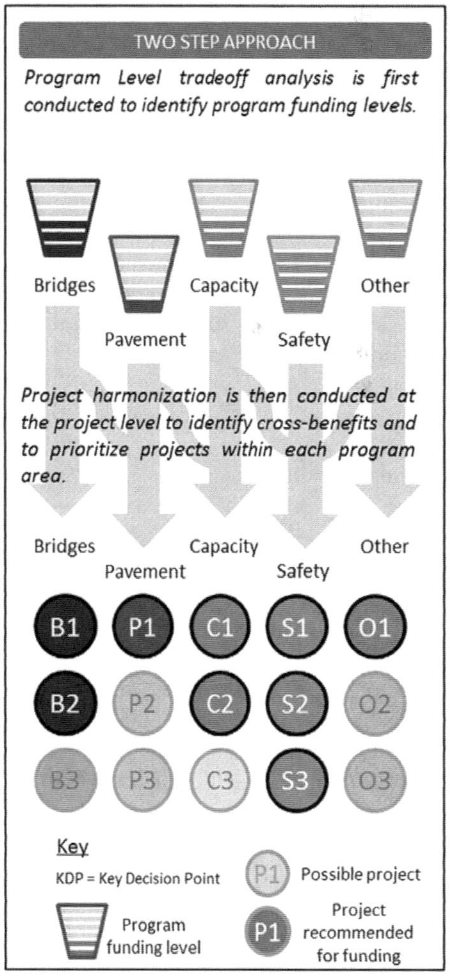

Figure 2. Recommended two-step approach to tradeoff analysis.

will not generate an optimal work program. Project proposals are asset, corridor, or region-level maintenance or capital tasks under consideration for the work program. The two-step approach assumes that a DOT has multiple avenues of generating project proposals, including the engineering judgment of staff, expressed regional preferences through stakeholder input and MPOs transport improvement plans, and output from asset management data systems. Recognizing that agencies can (and in many cases do) use all three of these approaches in tandem, the recommended two-step approach does not stipulate a single method for generating proposals, but is flexible to account for a combined approach. Additionally, the recommended approach emphasizes feedback from project selection and implementation to proposal generation. In general, programs should plan for a realistic level of funding, managers should develop enough projects to efficiently execute the largest plausible budget, but not so many that they establish unachievable performance targets. To do so, they must be provided with feedback from later phases in prior rounds of allocation. One additional recommendation is that proposals reflect program managers' understanding of efficient scope and scale. If combining projects realizes cost, schedule, or impact benefits (e.g., by performing work on a corridor all at once, or by repairing a bridge deck in concert with repaving) then proposals should be combined from the outset.

- Distribute funding among programs: The two-step approach adopts a performance-driven allocation process from the top–down approach, respecting statutory limitations and dedicated funding sources. The DOT builds response curves by program to relate system performance to funding level. The recommended approach then generates an ideal distribution scenario to maximize performance by either manually varying investment levels and reviewing performance in each case or by optimizing performance functions through linear scaling, monetization, or a value/utility function development process.
- Prioritize projects for inclusion in the work program: The two-step approach adopts the harmonization process from the bottom–up approach as one possible project selection method, but applies it within programs. In the harmonization process, project metrics are applied, as generally as possible, to ensure the consideration of co-benefits. Prioritized lists of projects within programs are reconciled with each other, and allocations are adjusted, if necessary, to ensure the most optimal possible work program. The recommended approach also respects established and effective project selection methodologies by program. Familiarity and confidence in existing practices may make them more efficient and effective than theoretical alternatives.
- Schedule the work program: The work program is scheduled by the DOT personnel using considerations that are meaningful to the DOT. These include the level of design progress of each project (e.g., how shovel-ready it is), as well as statutory requirements, availability of labor, equipment, materials, and contractors, acquisition timetables for materials and treatments, temperature and weather, geographic, economic, and demographic equity, regional priorities, and political realities, among others. Ideally, the DOT will develop a whole-life cost measure that reflects the timing of the work (i.e., "how much do we save by doing this now versus doing it next year, in five years, or in twenty years?").

In addition, the following recommendations for enhancing tradeoff analysis and cross-asset resource allocation processes were developed from the work reflected in this paper:

- Approaches to account for co-benefits of projects should be developed. These should reflect that reduced costs can be realized when two projects in close geographical proximity are conducted together.
- The assumption of utility independence with respect to managing several assets needs to be investigated. The current assumptions are that the utility gained from improving one asset do not depend on the condition of other assets, and this assumption has an impact on the mathematical formulation of the tradeoff analysis.
- A feedback loop is necessary to capture the effects of deferred maintenance on resource allocation decisions. For example, two assets may provide the same benefit, and thus are given equal preference when programming projects. However, if the potential increases in cost resulting from deferring maintenance on one asset are significantly higher than the deferred cost of the other asset, different preferences should be assigned to each asset.

Finally, resource allocation requires that the decisions made within an agency reflect the priorities established by the agency, and ultimately these priorities reflect the preferences of the travelling public. Therefore, approaches to resource allocation need to maintain flexibility to adapt to changing public preferences. Any approach to resource allocation must be able to capture the dynamic political and social preferences of modern times.

Author Contributions

J. Bryce, G. Rada, S. Van Hecke and J. Zissman contributed to the design and implementation of the research, to the analysis of the current state of TAM practice, to the development of the methodology and to the writing of the manuscript. All authors reviewed the results and approved the final version of the manuscript.

References

1. Cambridge Systematics, Inc. *NCHRP Report 632: An Asset Management Framework for the Interstate Highway System.* Transportation Research Board of the National Academies, Washington, D.C., 2009.
2. *Transportation Asset Management Guide.* AASHTO, Washington, D.C., 2002.
3. Cambridge Systematics, Inc. *NCHRP Report 551: Performance Measures and Targets for Transportation Asset Management.* Transportation Research Board of the National Academies, Washington, D.C., 2006.
4. AASHTO. *Transportation Asset Management Guide: A Focus on Implementation: Executive Summary.* FHWA-HIF-13-047. FHWA, U.S. Department of Transportation, Washington, D.C., 2013.
5. *MAP-21: Moving Ahead for Progress in the 21st Century.* FHWA, U.S. Department of Transportation, Washington, D.C., 2013. http://www.fhwa.dot.gov/map21/. Accessed February 17, 2016.
6. FHWA. *Executive Brief: Advancing a Transportation Asset Management Approach.* U.S. Department of Transportation, Washington, D.C., 2012. www.fhwa.dot.gov/asset/pubs/if12034.pdf. Accessed May 8, 2017.
7. Maggiore, M., and K. M. Ford. *NCHRP Report 806: Guide to Cross-Asset Resource Allocation and the Impact on Transportation System Performance.* Transportation Research Board of the National Academies, Washington, D.C., 2016.
8. Cambridge Systematics, Inc. *NCHRP Report 836: Performance Measures for Infrastructure Preservation.* Transportation Research Board of the National Academies, Washington, D.C., 2014.
9. Keeney, R., and H. Raiffa. *Decisions with Multiple Objectives: Preferences and Value Tradeoffs.* Cambridge University Press, Cambridge, MA, 1993.
10. Cambridge Systematics, Inc. *NCHRP Report 545: Analytical Tools for Asset Management.* Transportation Research Board of the National Academies, Washington, D.C., 2005.
11. Dehghani, M., F. Giustozzi, G. Flintsch, and M. Crispino. Cross-Asset Resource Allocation Framework for Achieving Performance Sustainability. *Transportation Research Record: Journal of the Transportation Research Board,* 2013. 2361: 16–24.
12. Maine DOT. *MaineDOT Asset Management – Highways.* Maine Department of Transportation, Augusta, ME, 2016. http://maine.gov/mdot/about/assets/hwy/. Accessed June 23, 2016.
13. FHWA. *Fixing America's Surface Transportation Act or "FAST Act".* U.S. Department of Transportation, Washington, D.C., 2017. https://www.fhwa.dot.gov/fastact/summary.cfm. Accessed March 13, 2017.

The Standing Committee on Transportation Asset Management (ABC40) peer-reviewed this paper (18-01200).

Article

TRR
JOURNAL OF THE TRANSPORTATION RESEARCH BOARD

Transportation Research Record
2018, Vol. 2672(44) 32–42

DOI: 10.1177/0361198118796069
journals.sagepub.com/home/trr
SAGE

Validation Protocol for a Pavement Management System

Theunis F. P. Henning[1], Elke Beca[2], and Gordon Hart[3]

Abstract

Being able to undertake more efficient maintenance and renewal planning of road networks has seen a number of pavement management systems (PMS) being developed between the 80s to mid-90s. During the initial years, there was a significant focus on substantiating the need and demonstrating the benefits of using these tools to assist in the investment planning for roads. Today, with substantial experience, and legislative support such as MAP-21 and PASS-55, the use of PMS in the asset management cycle of roads is a given. The challenge these days is to effectively use these tools and ensure robust outcomes are achieved from the overall PMS process. The success of any PMS is determined by a number of factors including having skilled resources, institutional support, and legislative and funding drivers demanding evidence-based forecasted investment needs. Asset managers also realize the importance of having technical robustness in the process to ensure likely outcomes. This paper documents the evolutionary development pathway of the New Zealand PMS that has now been in use across the entire country for more than 18 years. It shows how by always challenging the status quo it has resulted in consistently increasing the overall robustness of the system. In particular, it shows how different validation techniques have been used to improve the practicality and appropriateness of long-term forecasting capabilities.

Most countries have seen a rapid road-network expansion post World War II. An assumed pavement design life of approximately 30–40 years meant substantial renewal requirements starting from 1980 onwards. This renewal need has placed a significant planning onus on network engineers, which saw the development of a number of pavement management systems (PMS). Although the functionality and sophistication of the various applications differ significantly, asset managers soon realized that the successful implementation was not only a function of the system, but it also depended on a number of institutional issues and the available data.

New Zealand (NZ) has always been in a fortunate position of having a dedicated road fund that is managed by the NZ Transport Agency (NZTA). One of the benefits from a central funding agency is that it can encourage a consistent approach throughout the local councils. This drive towards consistency facilitated the implementation of a shared database and pavement deterioration modeling application. By working together, the economy of scale and commitment to make the system work saw some significant development work resulting from the 19 years of using the system.

Local councils and the NZTA secured a national site license for the dTIMS software during 1998. The full configuration of the system including pavement models and treatment logic was developed using a collective of New Zealand consultants and contractors. The collaboration between the consultants and contracting industry ensured not only early buy-in to the initiative but also allowed for the capturing of best practices from all the asset managers throughout the country. The knowledge base is extensive on PMS systems, modeling approaches, and successful implementations. However, what is not that widely documented is the longevity of the system and how it was improved and validated over time. This paper took a retrospective review of the system implemented in NZ and demonstrated how the system was validated to ensure it is providing robust outcomes.

The objective of this paper is to:

- Briefly introduce the development evolution of the PMS system and process

[1]Department of Civil and Environmental Engineering, The University of Auckland, Auckland, New Zealand
[2]Opus International, Tauranga, New Zealand
[3]New Zealand Transport Agency, Napier Hawke's Bay, New Zealand

Corresponding Author:
Address correspondence to Theunis F.P. Henning: t.henning@auckland.ac.nz

- Demonstrate the validation processes that have been used to test and improve the robustness of the system

Introduction to the NZ Pavement Management System

The NZ configuration/setup has been evolving over the past 18 years in order to more closely align to observed road performance through the use of the NZ Long-Term Pavement Performance (LTPP) data. Figure 1 illustrates the building blocks of the system consisting of the data module, forecasting elements and the decision logic and optimization. The system itself has been chosen for the flexibility in building a customized analyses approach such as models and treatment logic within an analytical framework. From the building blocks indicated in the figure, it is only the optimization process that was resident in the system; the remaining technical makeup is solely based on NZ practices.

The resulting model framework includes five, nationally customized, core condition models including:

- Cracking (initiation and progression for asphalt and chipseal surfaces)
- Raveling
- Rutting (progression and accelerated progression)
- Roughness
- Maintenance cost

Within the setup, these core models are supported by an array of coding with over 300 expressions and 150 analysis variables covering data management, forecasting capabilities, decision logic, and optimization as illustrated in the figure.

Critical to the success of any analysis is a skilled operator, with an in-depth understanding of both the core logic and condition model performance and with the ability to efficiently customize and calibrate the setup for local conditions. Therefore, the training and other development needs were as important as the technical solution. The following sections discuss the implementation philosophy in more detail.

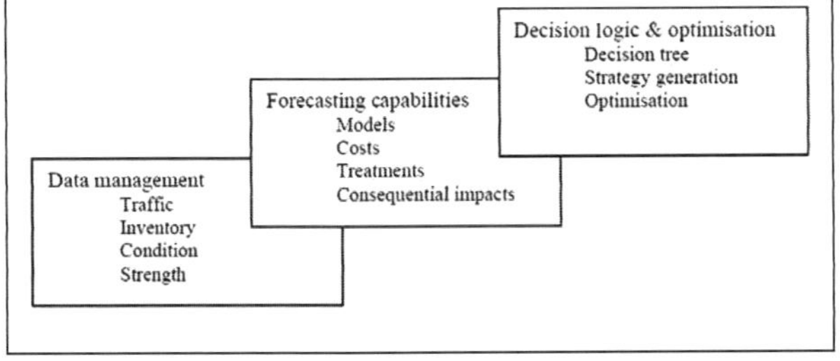

Figure 1. The components of the PMS (*1*).

Evolutionary Development Path

Initial Implementation

The initial implementation and some early lessons have been discussed in Hatcher and Henning (*1*). One of the greatest success factors for the NZ PMS implementation was the collaborative environment that underpinned the project from the start. Under a nationwide membership agreement, all parties contributed their experience and developments in an open-source environment. Significant development needs and software improvements were collated annually and prioritized according to addressing all parties' interest in the most efficient manner. The software vendor has been very responsive in addressing identified application improvements because most of the development needs were also identified as moving the international best practice of the application forward.

Another significant success factor that contributed to the overall sustainability of the system implementation was considering all peripheral business processes and institutional aspects that are as important as the technical aspects. Refer to Figure 2.

The figure shows an implementation strategy that acknowledges the status of skills, organizational understanding, and the state of data and deterioration models feeding into the system. A number of training sessions and a series of guidelines and manuals were released during the initial stages of the project. Wilson et al. (*2*) have described the data improvements to better cater for the deterioration models. The following sections consider some developments of the significant model and treatment logic aspects.

Pavement Model Development

In order to accelerate the system adoption, a decision was made to adopt the World Bank HDM-III mechanistic–empirical or deterministic models and analysis philosophy (*3*) as a starting point for the NZ implementation. The reasoning was that (1) the data and knowledge on the pavement performance in 1998 were lacking, and (2) the HDM-III models are transferable between environments and with calibration could be relatively robust in a short time frame. The intention was to develop data-driven models specifically applicable to NZ conditions. The LTPP was established in 2001 and has provided the research data for new model development, regional calibration of existing models, and data practices for more than 16 years (*4*).

A novel modeling approach that resulted from research on the LTPP program data was the use of

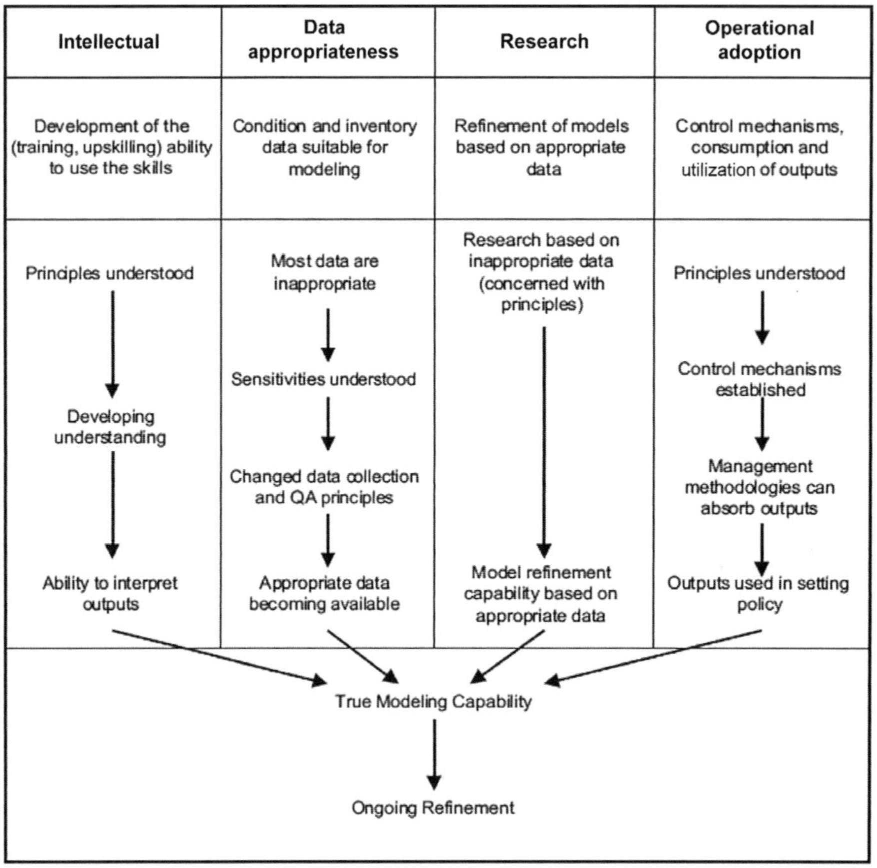

Figure 2. Development streams for the NZ PMS system (2).

continuous probabilistic models to forecast initiation of defects such as cracking and raveling. It is also used to forecast the point in time/inflection points of accelerated deterioration for the rutting model. These models yield the probability of a defect occurring given the dominant environmental, traffic and pavement strength characteristics (*5*, *6*). A later section provides the validation outcomes of the models.

Catering to Specific Road Types in the Configuration

Once some experience had been gained in the system, it became apparent that the deterioration of the roads in NZ is somewhat different to the HDM-III study area. More than 70% of NZ roads can be classified as low-volume roads (LVR) carrying less than 5,000 vehicles per day. At this traffic volume, the climate and geological interaction have a greater impact on road deterioration than just the traffic loading. Therefore, at lower-volume road spectrums, the deterioration models are less transferable between regions compared to higher volume roads. Figure 3 illustrates the difference in performance between LVR in NZ compared with the HDM model outcomes.

The figure suggests where the HDM models suggest a continuous deterioration as a result of traffic loading and environmental effects, NZ LVR would have very little deterioration over time, yet at a given period in time, the pavement will undergo accelerated deterioration and maintenance intervention would be required within months, normally within a year. The system was customized for this behavior by:

- Adjusting the model approach discussed in the previous sections
- Changing the maintenance intervention philosophy to focus more on preservation treatments such as resurfacing
- Using composite indices that not only account for observed defects but also include an aged-based deduct factor (*7*)

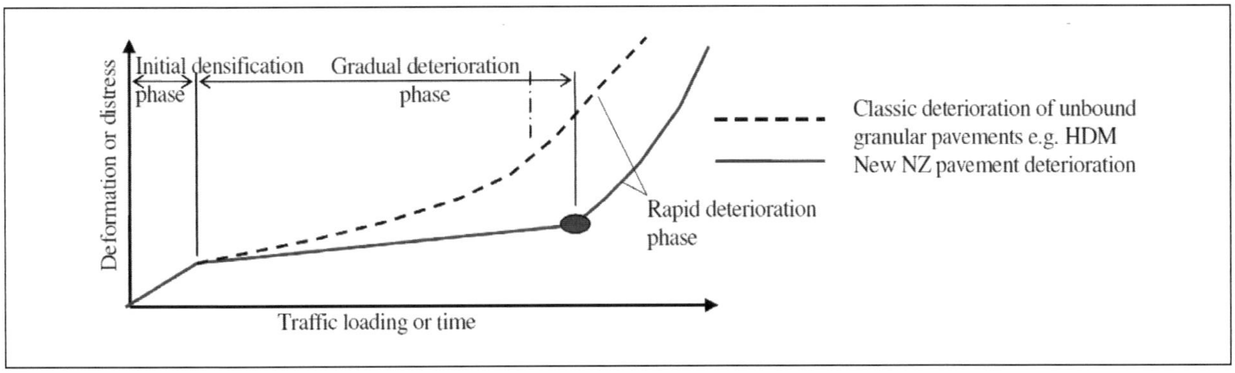

Figure 3. Typical deterioration of LVR in NZ (*1*).

Generic Calibration Process

Calibrating the condition models to a network/local level is an essential task undertaken at the outset of the analysis. Note that this step is taken each time a new analysis is made on a network. With the data improvements, the robustness of the models also improves. Once the calibration outcomes become stable for a given authority, the model robustness is only validated on a bi-annual cycle. Both the original HDM models and the newly developed models for local conditions are mechanistic–empirical. Therefore, a standard model format is being used for all networks with calibration factors allowing for incorporating local environment and geological conditions. As an example, the calibration process is illustrated below using the NZ rut progression model:

- First, the level of calibration is determined on the basis of the available data. Some regions have detailed calibration LTPP sections, and for others, network level data are being used.
- The models are then calibrated using the expressions provided in Equation 1 and Equation 2. Equation 1 returns the annual rut increment provided that the road section has not yet fully deteriorated into an accelerated rut progression.

$$\text{Annual Rut Increment} = \text{IF(Accelerated Rutting Initiated, 1.4, 1.0)*Calibration Factor*0.1*}\frac{\text{LOG10(MAX(10.0, Traffic ESA*365.0))}}{\text{Structural Number}} \quad (1)$$

$$\text{Calibration Factor} = \frac{\text{(Annual Increment*Structural Number)}}{\text{IF(Accelerated Rutting Initiated, 1.4, 1.0)*0.1*LOG10(MAX(10.0, Traffic ESA*365.0))}} \quad (2)$$

where

Accelerated rutting initiated = TRUE once rutting goes into the accelerated rut progression (refer to Figure 3)

Traffic ESA = standard average axle loading per day

Structural number = modified structural number as defined in Watanatada et al. (*3*)

Note that the historical condition data is aggregated to the defined level, cleaned to remove maintenance or renewal impacts, and the rate of condition change assessed.

A suite of tools has been developed to assist the NZ modeler in the calibration process and understanding of the model outcomes in using different calibration methods. One such tool, illustrated in Figure 4, provides an interactive dashboard where the user may adjust input variables, including calibration factors to plot model outcomes and assess the impacts visually. This tool assists the modeler to gain a better understanding of critical inputs to each condition model as well as demonstrating the impact that calibration factors have on condition models.

Validating the Results

Background National State Highway Analyses

In 2016, a robust least life-cycle cost analysis was performed over the full length of the NZ state highway net-

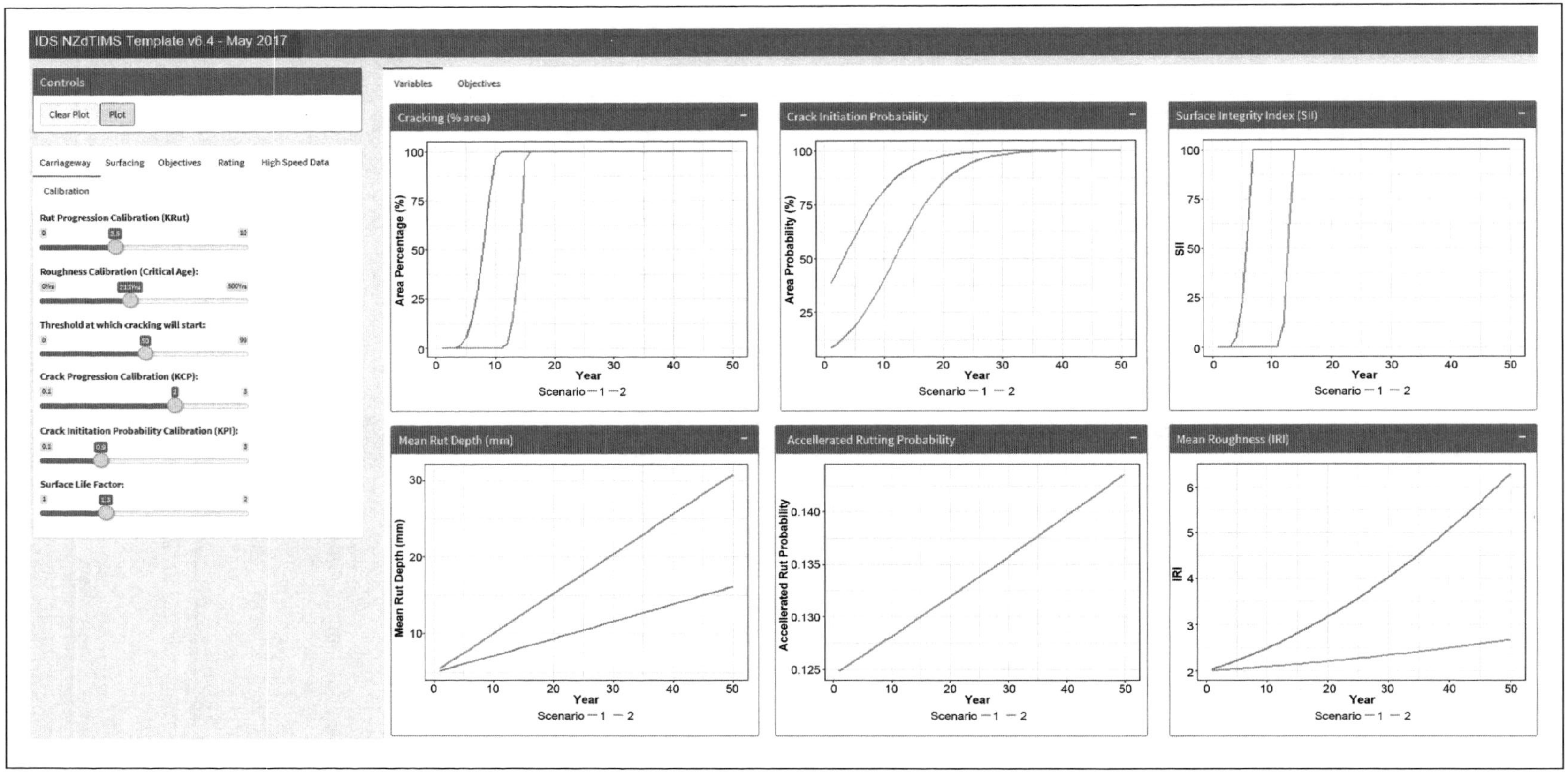

Figure 4. Screen shot from calibration tool.

work (approximately 23,000 lane km). The objective of the analysis was to support the 10 year investment request for the pavement asset to government funders, through evidence-based justification. Given the model's outcomes are being presented as such evidence, the funder also required an assessment of the robustness of the model through sufficient validation. The analysis drew upon innovative methods to calibrate, test, and validate the outcomes, providing a level of confidence in the outcomes unsurpassed. The following sections step through the technical details of these verification phases undertaken throughout the analysis.

Model Calibration

Prior to the analyses, a similar calibration process described in the previous section has been undertaken on the state highways. In addition to the LTPP initiative, NZ is fortunate to have a longstanding national data collection program across all state highways:

- Continuous high-speed laser profilometer data (rutting, roughness, texture, geometry, and skid resistance), 100% annual survey since 1996
- Visual condition rating data (surface, pavement, and shoulder defects) at 10% sample size annually between 1992 and 2014
- Falling weight deflectometer (FWD) at 100 meter drops, covering approximately 20% network every 5 years since 2001
- Continuous traffic speed deflectometer data (TSD) in 2015 and 2016

The availability of this wealth of continuous network data, matched with reliable construction history information enabled a higher level of calibration in this analysis. Calibration groups were defined based on geography/region (23 in total) and further split by functional group or road classifications (12 in total), resulting in 276 calibration groups. The rate of condition deterioration, following the latest major construction, was assessed for each modeled section and using these, a calibration factor then assessed for each calibration group.

Rolling Back the Clock

While local calibration has been completed prior to analyses, a further test was undertaken to prove the appropriate predictions from the models in the context of the overall system's logic. Retrospective analysis or "rolling back the clock" is a method used to test the entire forecasting logic and is an efficient method for post-analysis verification. This method involves rolling the entire dataset back a defined number of years, locking in the treatments which have actually occurred on the network over the defined period, running the analysis, and comparing the model-generated network condition with the actual network condition.

This exercise was completed as part of the 2016 NZ State Highway analysis to assess the predictive capability of the calibrated rutting and roughness condition models. A 10% sample of the national dataset (three network areas out of 23 total) was selected with input data rolled back by 5 years. All renewal treatments which had occurred on the network over the five-year period were identified and locked into the model, triggering a condition reset in the model in the actual retrospective treatment year. The model was run over the 5 years with the final year condition projections compared to actual measured condition values at project (section) level. Outcomes are shown in Figure 5.

The chart on the left shows the cumulative distribution of the three networks' roughness condition in the final year, with model and actual lines almost indistinguishable. This suggests at a network level, the predictive capability of the model of an acceptable robustness.

The chart on the right shows a correlation between model and actual data points at a section-length level. The R^2 value of 0.87 suggests a high correlation, providing confidence in the predictive capability of the prediction model even at a project level. The outliers are primarily represented by the sections where treatment has occurred, and the model has reset the condition at a fixed level. These outliers indicate either that the condition resets coded within the model are too low and would benefit review or the historical treatments coded into the retrospective analysis did not actually occur. As demonstrated in the example, the retrospective analysis is an effective exercise to not only verify the accuracy of predictive models but also the wider model logic such as condition resets.

Probabilistic Model

Historically, NZ has adopted a single view of pavement performance modeling based on a single deterministic framework as documented in this paper. While the condition models and decision logic have been evolving over the past two decades, the model framework itself has remained static, with all modelers nationwide operating on the same, single, base setup using the same PMS.

To test and challenge the outputs from this well-established framework, in 2016, a new setup, based on probabilistic principals was developed. The probabilistic setup was built around a Markovian transition probability matrix (TPM) framework, with the TPMs developed and calibrated using historical data from the full NZ State Highway dataset (*8*). Figure 6 illustrates the difference between the two model forms, both members of the empirical model family.

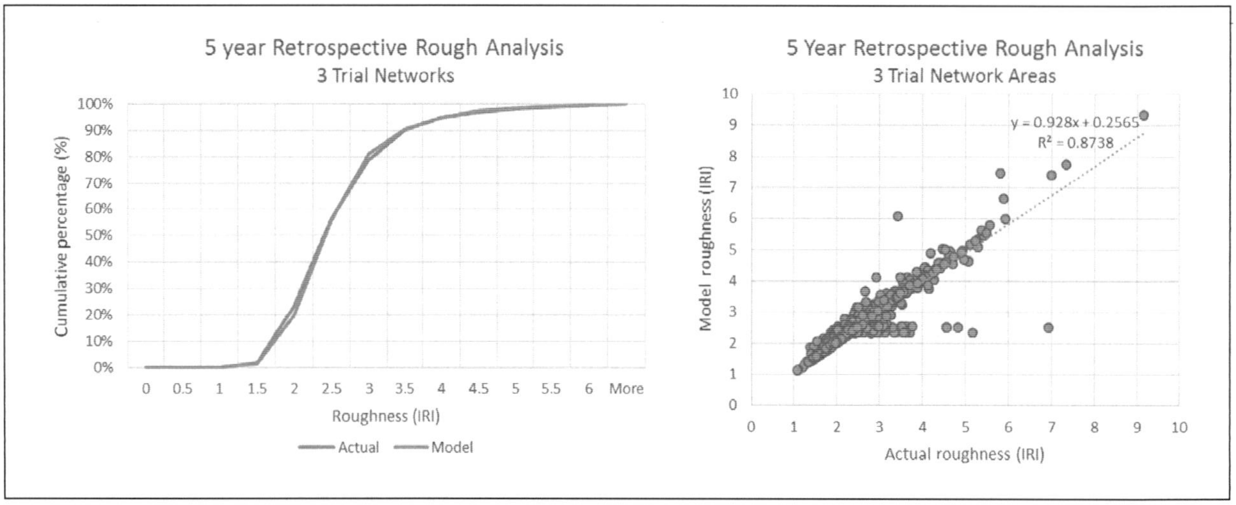

Figure 5. Comparing actual to modeled conditions for retrospective analysis.

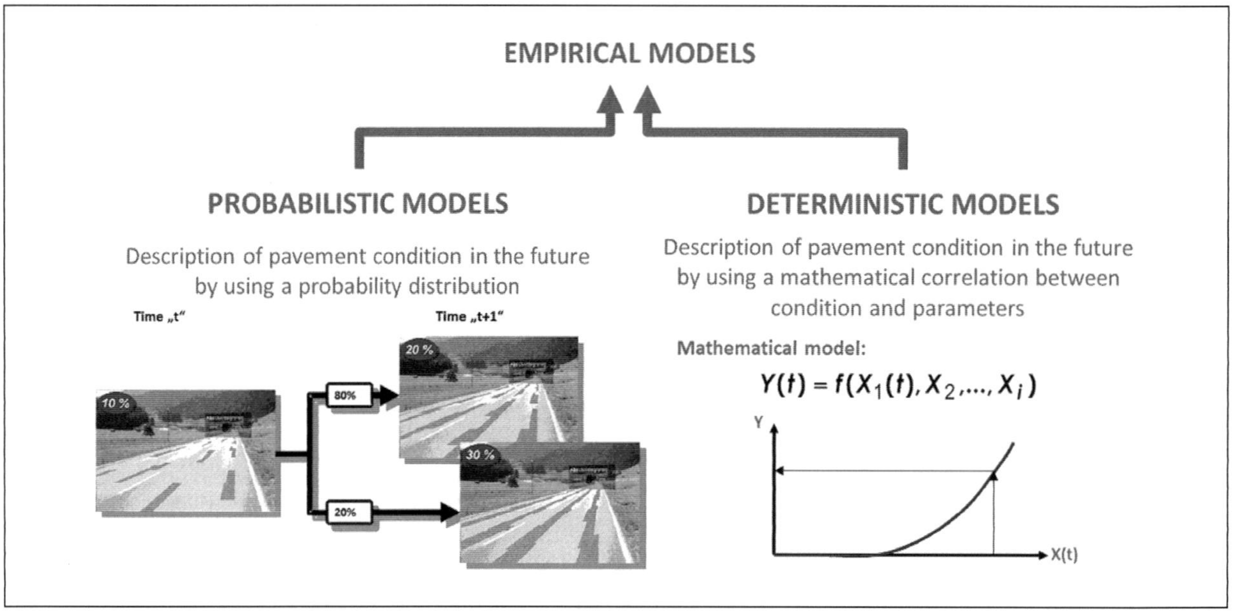

Figure 6. Graphical illustration of the difference in deterministic and probabilistic model approaches.

Development of the probabilistic models (TPMs), while simplistic with just three initial index-based models (Pavement Renewal, Surfacing Renewal, and Safety Renewal), was well-researched and models were subjected to rigorous validation and testing prior to adoption.

The approach adopted to challenge the validity of the existing NZ models was to code a like-for-like setup, where all components of the modeling framework were identical apart from the condition models themselves. The result was a second, standalone NZ roading model framework within the same PMS, with entirely different condition models driving investment recommendation. An entirely new view of the same landscape, would the two frameworks produce similar results to the question, What investment level is required to maintain the current steady state condition of the network?

To enable comparison, the optimization routines of both the deterministic and probabilistic PMS frameworks were adjusted to seek steady state condition or in

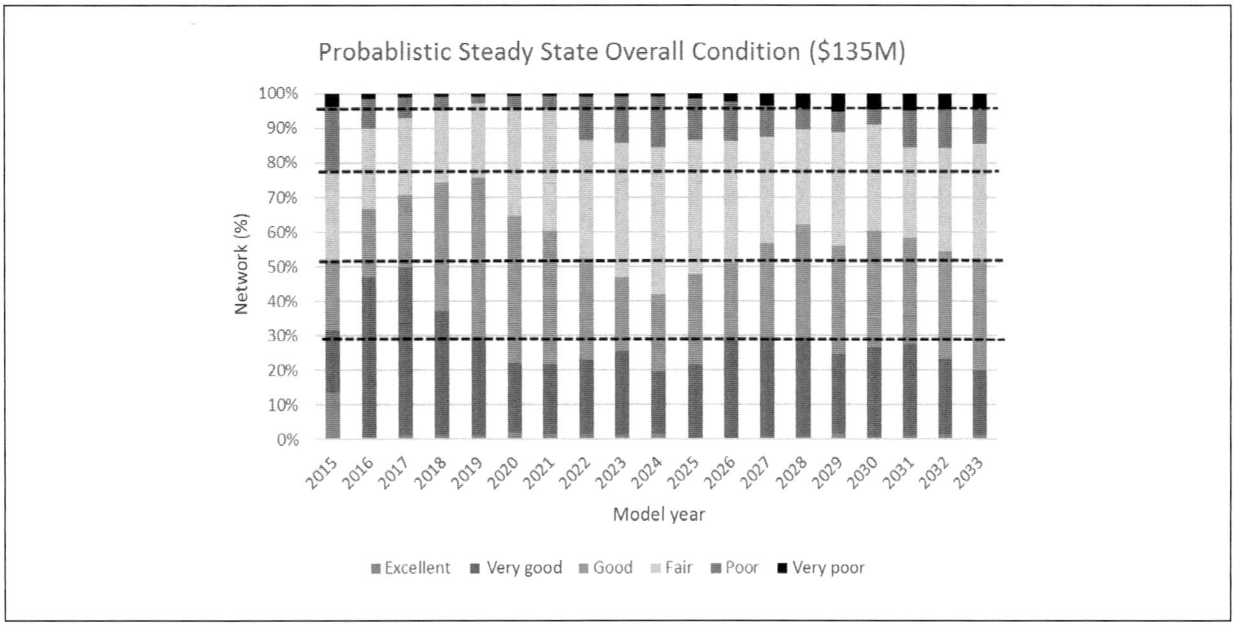

Figure 7. Condition forecast from Markov approach.

other words, maintain the current overall network condition (objective function). Five scenarios were tested under each framework over a range of 30% investment. The inherent cyclic nature of the Markovian TPMs was observed in the probabilistic model outputs, making it difficult to predict a continuous steady condition state as demonstrated in Figure 7.

The deterministic model predicted a stable condition outcome, providing a clear distinction between investment levels as demonstrated in Figure 8. Note, it is apparent in this figure that initially the model is tackling an observed "backlog" which boosts the Overall Condition Index in the first 2 years. Steady state has been identified where the profile stabilizes over the 3–20 year period.

Overall outputs from the two different model frameworks predicted steady state investment levels annually over a 20 year period within 5%–7% of each other. While the probabilistic framework is simplistic and has an enormous capacity to be further improved, this comparative assessment demonstrated that at a strategic investment level these two entirely disparate modeling approaches aligned, proving a further validation of the NZ model's predictive capability.

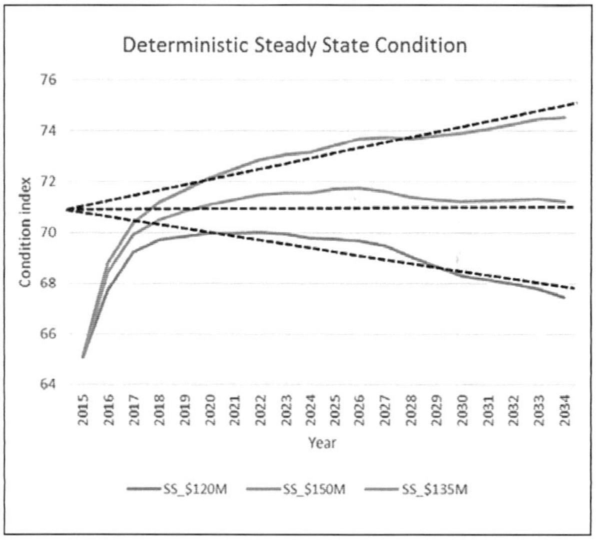

Figure 8. Condition forecast from deterministic approach.

Using Field Validation to Improve Forecasted Outcomes

Introduction to Field Validations—RAPT Reviews. The NZ Transport Agency utilizes a comprehensive field validation process for determining the maintenance and renewal program for up to three years. The so-called "RAPT reviews" (Review and Prioritisation Team) are undertaken across all regions and approve all resurfacing and rehabilitation projects that are identified as potential candidate projects for the following three years (*9*). The RAPT team consists of a number of representatives from the agencies' central planning team, regional staff and contractors responsible for the works program for a particular region. The team, traveling together in a bus, would visit all identified candidate sites to make a final

Table 1. Analyzing the Timing of Treatments According to the RAPT Review Process

Region	All treatments (%)			Rehabilitation (%)			Resurface (%)		
	Early	Right	Late	Early	Right	Late	Early	Right	Late
Region 1	8	90	2	14	86	0	18	79	3
Region 2	10	82	8	0	78	22	24	71	5
Region 3	19	81	0	27	73	0	30	66	4
Region 4	20	75	5	41	3	46	23	77	0
Region x	24	75	1	39	22	39	16	80	4

decision on the treatment strategy, being it to be included for construction or deferred and monitored for later years. The RAPT reviews are an integral part of the maintenance contracts (Network Outcome Contracts) that are in place to manage the road networks on the state highways. These contracts are performance-based contracts that are further described in Hunt and Hart (*10*).

The overall objectives of the RAPT reviews are:

- Improve consistency in performance across the Transport Agency's dispersed network.
- Develop the capability of team members.
- Develop a greater awareness of strategic focus that the Transport Agency wants to see embedded in forwarding works programs.
- Highlight asset management and work practice opportunities.
- Spread good ideas around the country (*9*).

The RAPT reviews take a number of information sources and program suggestions to inform their decision process. Some of the information sources are past condition performance, maintenance records, the existing 10 year program, the forecasted modeling results, safety performance, and existing contractors' programs. The existing draft program is then assessed according to the final decision process and reported for its appropriateness both in terms of timing and type of treatments (Table 1).

Incorporating RAPT into Modeling Process. Asset managers should be careful in using field validation to check on the robustness of modeling outcomes. The fundamental postulation of such a comparison suggests the field decision is always right. This raises the question, If the field decision is always right, why do you need to run a model in the first place? Field inspectors are capable of judging future performance for a short time only and, unlike the model, often make inconsistent decisions across a range of candidate sections. It has to be realized that: given that the modeling takes account of the full life-cycle cost and benefits; given that it optimizes the outcome for the entire network; and given that it "views" an individual road section in the context of the overall network, the answer from the model will most likely not be the same as the field inspections. The recommended approach in using field validation with a modeling approach is as follows:

- When the modeling is still in its infancy implementation stage, field validations are useful to identify inappropriate renewal treatments. Experience has proven that in 80% of cases inappropriate treatment forecasts can be attributed to data issues.
- Compare the network level predicted renewal quantities, costs, and condition profiles with actual historical achievement. An obvious test would be to assess the balance of predicted proactive and reactive investment levels against historical levels.
- Once the forecasting model has been sufficiently tested and refined, the emphasis shifts away from aligning the forecasted model and the final field decision, to acknowledge it may be different but focuses on reasons why the field decision and model outcomes show a different approach.

Validating the 2016 Model Outcome to the Field Program. At the onset of the Network Outcome Contracts, earlier modeling processes have been one of the core inputs to establishing the contractual renewal quantities or "baseline quantities" (*10*). Note that the baseline program was established using the deterioration model that was then validated using the RAPT process described in the previous section. A critical outcome from the process was a significant reduction of the renewal program compared to historical levels. With the 2016 analysis, it was, therefore, important to assess the model outcome in comparison with both the previous analysis (baseline quantities) and with the contractors' programs. The aim was, therefore, to establish whether:

- The network needs are still addressed in a sustainable manner.

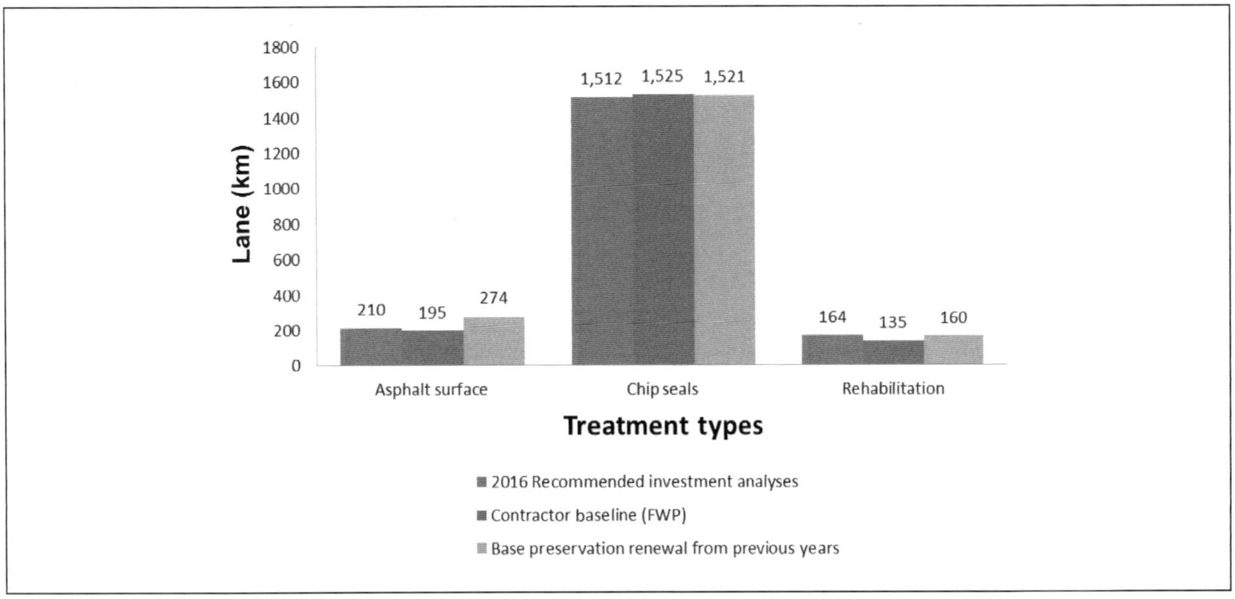

Figure 9. Comparing the model outcomes to existing programs.

- What the impact of the three-year program would be on the long-term performance and investment needs for the state highway.

Therefore, while the earlier analyses and RAPT review validated a program of works, the 2016 analyses' aim was to establish a long-term investment strategy. For this reason, a network-wide quantity outcome comparison was deemed more appropriate. Figure 9 shows the relative quantities between the model outcome, the contractors' baseline and analyzed outcomes from previous years. The high level of agreement between the programs not only sufficiently validated the 2016 analyses outcomes but further gave the agency comfort in terms of the appropriateness of their current approach to the Network Outcome Contracts.

Summary and Conclusion

This paper has documented the validation procedure undertaken on a PMS system adapted for NZ. Whereas many publications document the successful implementation of systems, this paper took a retrospective view of a system that has been used for more than 18 years. The paper has demonstrated a number of success factors that contributed towards the overall sustainability of the system. This is one of the outstanding aspects of the collaborative environment of the continuous technical development and improvements by modeling experts from a number of councils, consultants, and contractors.

Because the NZ PMS are being used in the determination of long-term funding requirements and applications by road agencies, the robustness of the system was of utmost importance. The paper has shown a number of validations processes used for the latest long-term investment modeling for the national state highways. This validation tested the model approach from any possible viewpoint that included:

- Calibration of deterioration models using recognized best practice. In this instance the calibration dataset included detail data from LTPP sites and some network data.
- In addition to the direct model calibration, the overall forecasting aspects of the system were tested by retrospectively modeling (rolling back the clock). A comparison was undertaken between the current condition to the five-year forecast from the system using condition data from five years ago and loading the actual maintenance and renewals program from the past five years.
- The system's modeling approach, originally based on the World Bank HDM-III modeling philosophy, was tested by undertaking parallel model analyses based on a Markovian probabilistic approach that uses condition transition matrices.
- Lastly, the system forecasted maintenance quantities were also compared to prior years accepted renewal and maintenance programs that were field validated.

A positive outcome was achieved through the validation process, not only proving beyond doubt that the system was robust within accepted confidence levels, but it also provided confidence to the agency that the current maintenance investment was at appropriate levels.

It is recognized that many agencies would not be in a position to complete the full validation processed presented in this paper. These agencies could consider parts of the suggested validation process to refine specific areas of concerns within their own systems.

The main lesson from the extensive testing regime of the PMS was that the validation ultimately resulted in the significant improvement of the robustness over the years. These improvements included data-quality improvements, refinements of forecasting models and improving the overall treatment logic of the system.

Acknowledgments

The authors would like to acknowledge the following organizations for their financial contribution to the material presented in this paper: New Zealand Transport Agency; all consortium members of IDS (Infrastructure Decision Support); Ola Shahin, developer of calibration tool; and Deighton Associates Ltd software providers and development partners in NZ.

Author Contributions

The authors confirm contribution to the paper as follows: study conception and design: GH, TH, EB; data collection: EB; analysis and interpretation of results: EB, TH, GH; draft manuscript preparation: TH, EB, GH. All authors reviewed the results and approved the final version of the manuscript.

References

1. Hatcher, W., and T. F. P. Henning. Lessons Learnt: NZ National Pavement Performance Model Implementation. *Proc., 7th International Conference on Managing Pavement Assets*, Calgary, AB, Canada, TRB Committee on AFD10 Pavement Management Systems, 2008.
2. Wilson, D. J., N. K. Pradhan, and T. F. P. Henning. The Appropriateness of Data in a Pavement Management System: The New Zealand Experience. *Road and Transport Research*, Vol. 11, No. 2, 2002, pp. 14–28.
3. Watanatada, T., C. G. Harral, W. D. O. Paterson, A. M. Dhareshwar, A. Bhandari, and K. Tsunokawa. *The Highway Design and Maintenance Standards Model.* World Bank, The John Hopkins University Press, Baltimore, MD, 1987.
4. Henning, T. F. P., S. B. Costello, R. C. M. Dunn, C. C. Parkman, and G. Hart. The Establishment of a Long-Term Pavement Performance Study on the New Zealand State Highway Network. *Road and Transport Research*, Vol. 13, No. 2, 2004, pp. 17–32.
5. Henning, T. F. P., and D. C. Roux. A Probabilistic Approach for Modelling Deterioration of Asphalt Surfaces. *Journal of the South African Institution of Civil Engineering*, Vol. 54, No. 2, 2012, pp. 36–44.
6. Henning, T., R. Dunn, S. Costello, and C. Parkman. A New Approach for Modelling Rutting on the New Zealand State Highways. *Road & Transport Research: A Journal of Australian and New Zealand Research and Practice*, Vol. 18, No. 1, 2009, pp. 3–18.
7. Fawcett, G., T. Henning, N. Pradhan, and M. Riley. The Use of Composite Indices as Resurfacing Triggers. *Proc., 5th International Conference on Managing Pavements*, Seattle, WA, TRB Committee AFD10 on Pavement Management Systems, 2001.
8. Patrick, J., E. Beca, and J. McLean. *Development of a Stochastic Model for Network Cost: Pavements.* Opus Research Report 14-521193.00.00. Wellington, New Zealand, 2014.
9. Road Efficiency Group. *NZ Transport Agency Programme Review Process.* Case Study 2013_01. Best Practice Asset Management Working Group, 2013. https://www.nzta.govt.nz/assets/Road-Efficiency-Group/docs/best-practice-case-study-summary-2013.pdf. Accessed November 2013.
10. Hunt, S., and G. Hart. Holistic Approach to Asset Management Delivery to Achieve Sustainable Efficiencies. Presented at WCPAM (World Conference on Pavement and Asset Management), Bavino, Italy, 2017.

The Standing Committee on Transportation Asset Management (ABC40) peer-reviewed this paper (18-01863).

TRR
JOURNAL OF THE TRANSPORTATION RESEARCH BOARD

Article

Transportation Research Record
2018, Vol. 2672(44) 43–54

DOI: 10.1177/0361198118790624
journals.sagepub.com/home/trr
SSAGE

Automatic Prediction of Maintenance Intervention Types in Roads using Machine Learning and Historical Records

F. J. Morales[1], A. Reyes[1], N. Caceres[1], L. Romero[1], and F. G. Benitez[1]

Abstract

A methodology to support and automate the prediction of maintenance intervention alerts in transport linear-asset infrastructures can greatly aid maintenance planning and management. This paper proposes a methodology combining the current and predicted conditions of the assets, and unit components of the infrastructure, with operational and historical maintenance data, to derive information about maintenance interventions needed to avoid later severe degradation. By means of data analytics and machine learning techniques, the proposed methodology generates a prioritized listing, ranked on severity levels, corresponding to the pre-alerts and alerts generated for all assets of the transport infrastructure. The methodology is applied and tested in a real case consisting of a road network with different section classes. The analysis of the results shows that the algorithms and tools developed have good predicting capabilities.

Research on linear-asset management (LAM) has attracted considerable attention in the last two decades; seminal publications have focused on specific types of linear assets, and important advances have been achieved in the management of railways (*1*, *2*), roads (*3*), cables (*4*), and pipelines (*5–8*). There are currently several enterprise asset management (EAM) tools dedicated to LAM (*9–11*) and also adapted general purpose EAM tools (*12*, *13*) on the market to address LAM issues. These lack predictive capabilities, however, or have very basic decision-making capabilities, even though these are strategic objectives for most of them.

It is worth remarking that most administrative bodies for transport infrastructure in developed countries are aware of the importance of reliable prediction tools in order to minimize downtime and to optimize the resources devoted to maintenance. The Federal Highway Administration has demonstrated that accurate knowledge of asset condition, along with proper management and analysis of gathered data, could lead to more effective maintenance planning, thus extending the life span of assets (e.g. pavements) by three to 10 years—an increase of between 12% and 20% (*14*, *15*). In the railway sector, the life span of ageing infrastructure can be extended by up to 25% through asset condition monitoring, prediction, and decision-making support (*16*).

Transport infrastructure maintenance alerts are customarily triggered when a fixed threshold, as defined by technical standards, is surpassed. These thresholds are prescribed by the corresponding infrastructure administration/regulator, based on knowledge accumulated from prolonged analysis of the adequacy and condition of the assets. The thresholds used are conservative in nature, thus guaranteeing the safety, integrity, and performance of the asset as a part of the system it works for. However, some alerts are only triggered by the appearance of failures, faults, or malfunctions, which in many cases have not been detected in advance due to a lack of awareness concerning explanatory features—thus the failure can be ruled a result of the absence of sufficient supervising/monitoring actions. In response to these cases, new explanatory features are considered and incorporated in the indexes to be monitored and measured. Ideally, every failure could be comprehensively explained using a fixed set of measurable features, however, this is unlikely due to the wide diversity of asset typologies.

At present, systems and procedures which identify critical explanatory features using data mining and analytics are being broadly investigated and developed. A structured database of historical maintenance

[1]Transportation Engineering, Faculty of Engineering, University of Seville, Seville, Spain

Corresponding Author:
Address correspondence to F. G. Benitez: benitez@us.es

interventions, with quantified information, paves the way for techniques which can infer "rules" that correlate failures. This in turn enables the creation of tools for either fully or semi-automatic decision making. Diverse methodologies have been used to find a way to infer those rules, most of them in the orbit of artificial intelligence (AI) and machine learning (ML), a field of computer science and engineering concerned with the computational understanding of what is commonly called intelligent behavior (*17*). An elaborate explanation concerning AI technique and discussion regarding other techniques can be found in the literature (*18*, *19*), including works specific to the LAM (e.g. railways) (*20–22*). In the road inspection field, studies in pattern recognition techniques and AI methods (*20*, *23*, *24*) show the capacity for fast outcomes with minimal computing requirements.

The field of maintenance alert prediction (*25*, *26*) has evolved hand in hand with the concept of reliability. Current practices focus on diagnosis as one of the main practical applications of AI systems, aimed at the prediction of faults or conditions in specific asset components before they evolve. The same is true of the evolution of transport infrastructure maintenance (e.g. road pavement conditions, geometry integrity). During the last two decades, an increasing number of studies have been published on the application of ML to LAM systems (*27–30*). In all cases, the common factor for this progression lies in the increasing availability of data, captured from auscultation/monitoring activities and campaigns; ML techniques promote the concept of learning from data, and facilitate the extraction of patterns and trends—letting the data "speak for itself." In recent years, efforts have been focused on improving the prediction capabilities of LAM systems (*31*). At present, all data-inferring-based disciplines (i.e. ML, data mining, statistics, big data) are working in the same vein; there is a remarkable amount of commercial software which has evolved from statistical origins (*32*). Despite this, the field is still in a state of rapid evolution.

The current paper presents methodologies, approaches, and models used for triggering alerts when there is a need for maintenance interventions, whether corrective, preventive, or predictive, associated with LAM of road infrastructures. The estimated alerts are assessed according to information provided by a decision-making tool, based on the forecast evolution of the condition of an asset, as reflected by relevant explanatory features, combined with a database of historical interventions. The output of this tool will tag each estimated alert with a level of severity, the estimated intervention, and their associated probability of occurrence. The final purpose is to provide a procedure for managing all active and predicted alerts, thus optimizing LAM maintenance operations.

Methodology Framework

The framework for predicting alerts as proposed in this research is outlined in Figure 1, where the different modules, interactions, inputs, and outputs are shown. Each module consists of techniques, methodologies, algorithms,

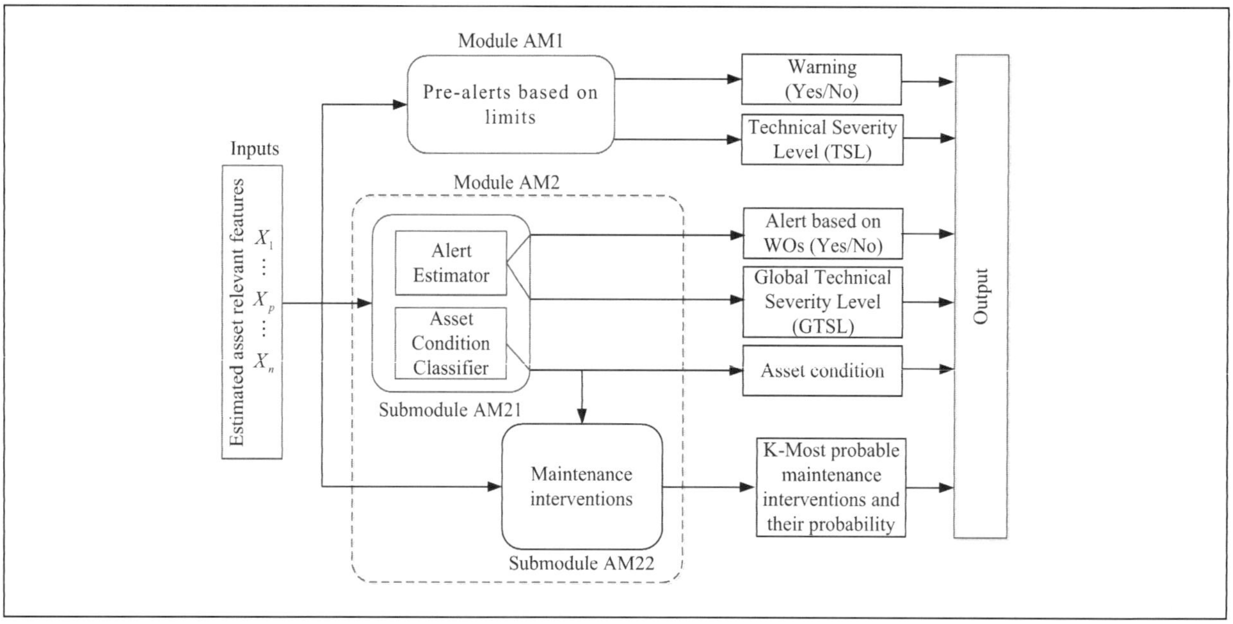

Figure 1. Methodology framework.

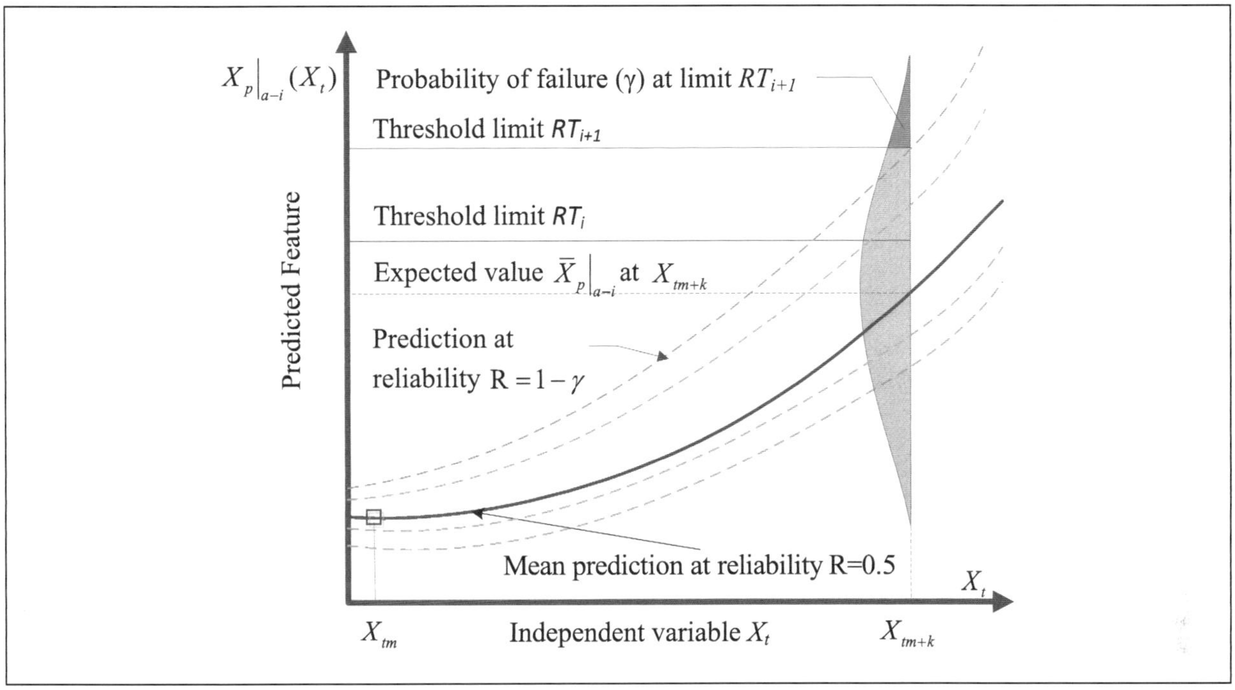

Figure 2. Asset condition prediction based on feature X_p, function of independent feature X_t.

and models, as explained below. The general objective pursued in this paper focuses on four main goals:

(a) Detecting pre-alerts and their probabilities, based on predefined threshold limits for the estimated values of the explanatory features of any specific asset, in predefined scenarios.
(b) Determining and triggering alerts, based on the predictions of supervised ML techniques and models which have been trained using the historical interventions database and the condition assessment conducted prior to carrying out intervention.
(c) Evaluating technical severity levels (TSL) of pre-alerts, and issuing a global technical severity level (GTSL) for each estimated alert.
(d) Estimation of the required intervention type for the triggered alert, along with a ranking sorted in descending order of probability of occurrence.

The developed methodologies implemented in each module are described below.

A Framework for Estimating Pre-Alerts Based on Feature Limits

The feature-based procedure (module AM1) is built on statistical techniques aimed at estimating the condition of specific explanatory asset features; for example, the International Roughness Index (IRI). The patterns found in the evolution of the condition of these features characterize not only the current condition of the asset, but also forecast future conditions, with their associated uncertainties. A description of how the inputs were processed follows.

Figure 2 depicts a sample of the evolution of the feature in a hypothetical scenario. It shows how the asset condition is estimated to evolve, characterized by a generic explanatory feature *p-th*, $X_p|_{a-i}$, of a particular road-segment asset, *a-i-th*, as a function of one or several other independent variables X_t (e.g. time, accumulated load). The most probable value is identified by the bold solid line of a fan of five evolving patterns, which correspond to probabilities with varied statistical reliability (e.g. $1 - \gamma$,0.5).

Two scenarios, X_{tm+k} and X_{tm}, are marked on the x-axis. The rightmost vertical cross-section line stands for the value of the independent variable X_t in scenario X_{tm+k}; in scenario X_{tm}, X_t is pinpointed by the square dot. Note that the dot might not lie on the bold solid line due to potential divergence between the predicted and the real values of the feature. The two uppermost horizontal lines identify two reference threshold limits (RT), denoted by RT_i and RT_{i+1}, set by the relevant standards for design/quality/safety parameters (e.g. normal limit L_N, exceptional limit L_E). The lower horizontal broken line identifies the expected value (i.e. with probability

0.5) at scenario X_{tm+k}. As defined by applicable criteria (e.g. European Standard, Road Administration Standard, Infrastructure Maintenance Managerial Body), the condition of the asset will be quantified in regard to the proximity of forecast values from the threshold limits. The probabilities of the estimated condition values are valuable, as they allow an assessment of the severity of the asset in each forecast scenario.

Figure 2 represents the estimated value of a feature with a known probability distribution. A pre-alert is generated when the condition of the asset, as identified by its explanatory features, surpasses a pre-set threshold under a specific probability in a forecast scenario. Any triggered pre-alert is then quantified; a TSL is associated with each explanatory feature, defined according to the distance from the predicted value to the pre-set limits (e.g. RT_i or RT_{i+1}). Both a TSL and the associated degree of uncertainty (γ) are defined; the straightforward evaluation of that distance can follow a multiplicity of criteria, as predefined by the maintenance managerial body (MMB).

Taking as an example the IRI of a road segment as X_p and assuming that the pre-alert will be triggered when a limit prescribed by a specific criterion (e.g. Standards) is surpassed, according to Equation 1:

$$P(X_p > RT_{i+1}) \geq \gamma \rightarrow \text{Pre-alert} \quad (1)$$

where the expression stands for the IRI surpassing a threshold with a probability larger or equal than γ. Depending on the value of γ (defined by the MMB), a pre-alert may be either considered or discarded even when the mean value of the feature is the same. The value γ is a means to demand a higher reliability for features considered more relevant to maintenance activities. A feature that is more critical, and thus only a low possibility of failure is allowed, will successfully issue a pre-alert at a lower γ value. In this case, $1-\gamma$ quantifies the reliability, thus, looking at Figure 2, Equation 2 is rewritten as:

$$\text{Prediction (at reliability} 1-\gamma) \geq RT_{i+1} \rightarrow \text{Pre-Alert} \quad (2)$$

So far, a pre-alert is triggered when the probability of failure exceeds the limit γ. For generalization purposes, a parameter α is introduced into the methodology. For instance, the MMB may consider (based on experience) that the fact of exceeding a reference threshold is not a strong reason to trigger the pre-alert, so the MMB may decide to introduce an additional gap (α) to trigger the pre-alert. The pre-alert has an associated TSL, so, when the TSL is greater than a predefined parameter α, the alert is triggered. An example definition for TSL_γ is presented in Equation 3:

$$\begin{aligned} &TSL_\gamma = \text{Prediction (at reliability} 1-\gamma) - RT_{i+1} \\ &\textit{If } TSL_\gamma \geq \alpha \rightarrow \text{Pre-alert} \end{aligned} \quad (3)$$

wherein, setting α to 0, the overall expression returns to Equation 1.

This example shows that, with generic values of TSL and α, it is possible to define a general pre-alert criterion; in practice, module AM1 embodies different strategies.

A Framework for Estimating Alerts and Interventions Based on Historical Work Orders

In cases where clear rules cannot be established with explanatory features—for example, due to variability in presumed-similar assets and endogenous and exogenous conditions—ML models, based on inferring knowledge from existing historical data, may pave the way to extracting the hidden know-how. This context is dealt with in module AM2 (Figure 1).

With the approach developed here, alerts are inferred by correlating the estimated values of the explanatory features, in a future scenario of interest for the MMB, with the recorded information stored in the historical intervention database. The relevant data stored in the repository contains, at least: (a) the intervention type, (b) the corresponding values of the explanatory features prior to the intervention, and (c) the maintenance team's subjective assessment just before the intervention. This means that alerts can be triggered not just by comparing estimated TSLs with pre-set thresholds, but also by using hidden, non-explicit information which may realize the needed intervention carried out in past cases. Through a training process, the hidden information (i.e. non-explicit thresholds) is extracted, to be used to estimate alerts. This procedure is implemented in two different functional submodules, based on distinct approaches. The first (AM21) is devoted to detecting whether an alert will take place, its estimated GTSL, and the estimated asset condition (using, as inputs, the forecast values of the explanatory features of the asset, with a pre-set reliability value γ). The methodology is built on a set of supervised ML algorithms. The second submodule (AM22) estimates the most probable intervention type, also providing the probability of this estimate.

The GTSL of an asset is derived from the TSLs of the corresponding explanatory features. This global index cannot be referred to any current threshold, as it is hidden, non-explicit information, rated in absolute value. The basic steps to obtain this GTSL are as follows:

(a) Evaluate the TSLs of each individual explanatory feature involved (TSL_1, ..., TSL_n). These can be calculated in terms of absolute values, implying that each TSL references the value of the corresponding feature, X_p, or in terms of relative values, comparable to prescribed threshold limits.

(b) Normalize the previous TSLs (TSLN) to refer all values to the same scale, in terms of their relative values: ($TSLN_1$, ..., $TSLN_n$). This normalization depends on the nature of the feature, and is subject to external criteria as defined by the MMB; for instance, the case of crocodile cracking—where the maximum value is 100—a normalizing criterion could be referred to this value.

(c) Weight each individual feature: $X_1 : \omega_1, \ldots, X_p : \omega_p, \ldots, X_n : \omega_n$, subject to $\sum_{i=1}^{n} \omega_i = 1$.

(d) Compute the normalized GTSL (GTSLN): $GTSLN = \sum_{i=1}^{n} \omega_i \cdot TSLN_i$.

At this point it is worth noting that, in most cases, intervention types are not directly associated with the asset condition that resulted in triggering a pre-alert, but influenced by other factors; for instance, two assets in the same condition could have intervention types depending on external factors (e.g. budget, available machinery, merging criteria to fix different sections belonging to the same road segment, maintenance policies, time opportunity). Due to this fact, the information to be used in module AM22 should be the proposed intervention to be conducted based on technical criteria, and not the actual intervention carried out based on other considerations. Failure to follow this rule will have negative effects regarding the training of ML algorithms and consequently their estimating capabilities.

Supervised ML Approaches for Estimating Maintenance Alerts and Asset Condition

In order to begin the process of estimating alerts, the first stage of the ML model (submodule AM21) involves a training process with the adequate information. The data repository used is a database of historical interventions regarding the monitored infrastructure case study. This database also contains various information regarding the subjective assessment of the condition of an asset, as inspected by the MMB team prior to intervention, in particular, the assessment of each individual explanatory feature (SX_p), each combined explanatory feature (CX), and a global evaluation of the condition of the whole asset (G). The predicted asset condition is an estimation of the subjective assessment, and the estimated global asset condition can be interpreted as the predicted condition of the asset when regarding the simultaneous contributions of all the explanatory features.

Other relevant information includes:

- the measurements carried out on the analyzed linear-asset segment (prior to intervention), associated with physical explanatory features
- endogenous and exogenous variables related to the evolution of the condition of the asset

The developed methodology of AM21 is split into two parallel prediction blocks, as shown in Figure 1. The *Alert Estimator* block is used to trigger alerts using a classifier, which simply correlates the values of the explanatory features with the requirement for maintenance. The target is a binary variable informing whether maintenance was needed or not (1/0 as Yes/No). The training methodology consists of four separate automatic binary classification ML models: decision tree (DT), *k*-nearest neighbor (KNN), support vector machine (SVM), and artificial neural network (ANN). The second block, *Asset Condition Classifier*, correlates the measured feature values with subjective evaluations of the asset condition as recorded by the MMB team. This way, the system "learns" from the practical experience of MMB and, when a new value is input, the model predicts the asset condition. According to this, the inputs used for training are the feature values, and the target variables are the subjective evaluations.

Once the system is initialized and the ML models are trained, the Alert Estimator is able to detect the road segments where maintenance will be required in a specific future scenario; the necessary inputs are the forecast values of explanatory features, $\hat{X} \equiv [\hat{X}_1, \ldots, \hat{X}_p, \ldots, \hat{X}_n]^T$, the outcome being the requirement (or not) for maintenance; where an intervention is forecast, an alert will be triggered.

Supervised and Unsupervised ML Approaches for Estimating Maintenance Types. The database of historical interventions provides data from the actual interventions carried out within a specific road network. In most cases, the reported information corresponds to the precise operations needed to recover the assets' functionalities, for example, interventions on detected faults/failures by either corrective/preventive actuations or predictive policies; in other cases, the conducted interventions followed other strategic considerations. It is possible to estimate the most probable intervention type in a given scenario, according to the know-how contained in the historical database, by correlating the estimated values of the relevant explanatory features, $\hat{X}$, and the forecast subjective state condition ($S\hat{X}, C\hat{X}, \hat{G}$) (provided by the second block of submodule AM21), versus similar asset samples reported in the historical database.

The prediction of the type of intervention is based on a two-phase unsupervised (automatic) ML scheme, carried out by the second submodule AM22 (Figure 1). This submodule determines a hierarchical listing of the most probable interventions needed, and their probabilities of occurrence, via training based on the historical database.

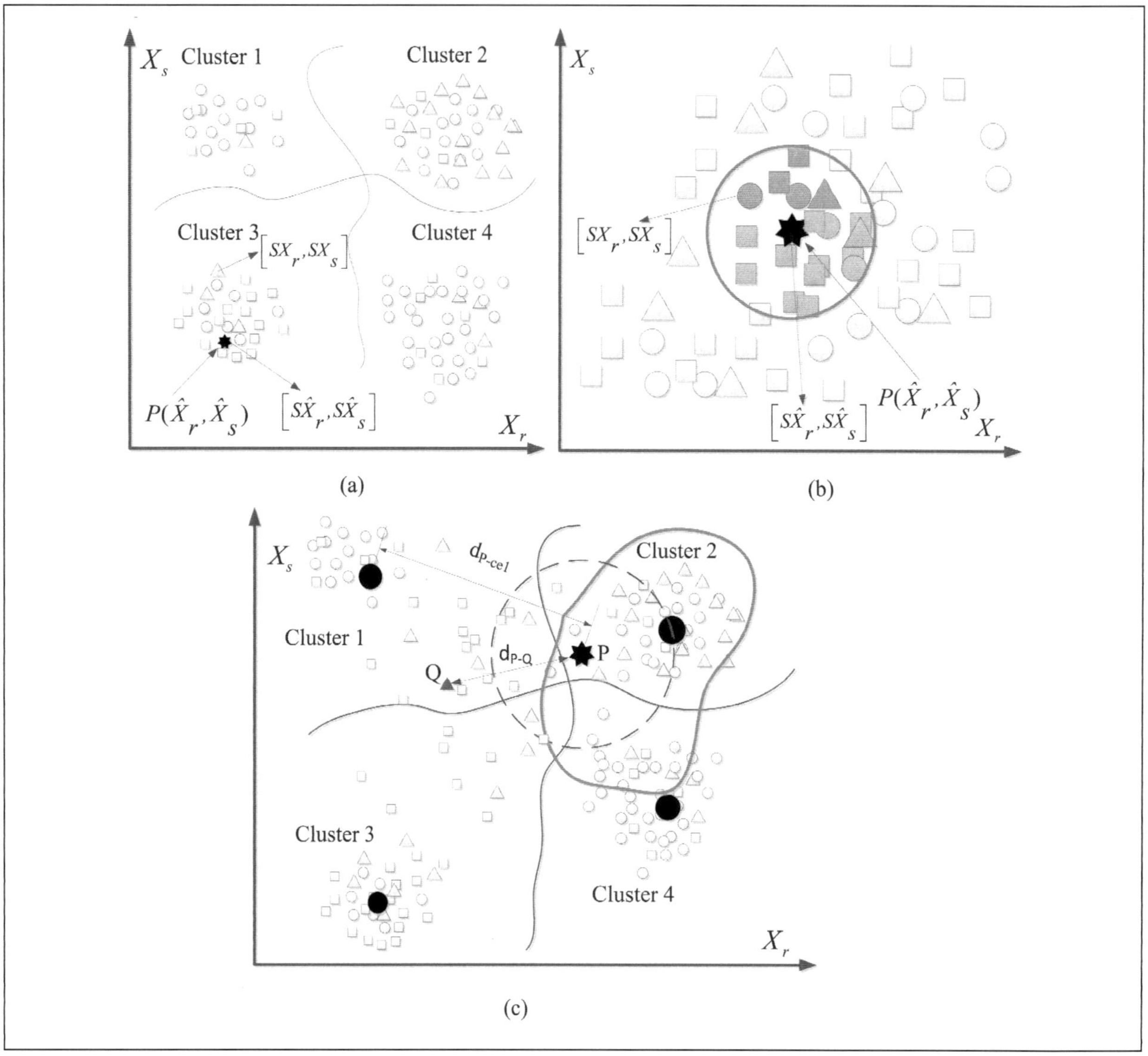

Figure 3. Prediction of intervention type and probability: (*a*) clustering from historical state condition; (*b*) KNN plus know-how approach; (*c*) fusion model.

The first phase is based on a k-clustering technique; the second phase, on a KNN technique from the outcome inferred by the Asset Condition Classifier block in AM21 (Figure 1) from a supervised ML (SML) technique. A description of the techniques used and methodologies implemented follows.

Unsupervised Clustering of Historical Data. The available historical data is grouped into different more or less well-defined clusters (e.g. L), each embodying samples in a similar condition (i.e. similar values for all explanatory features). The clustering is carried out taking into account just the asset condition (i.e. some value of features, X), without using the actual conducted intervention. Therefore, the same cluster could contain different intervention types, as shown in Figure 3*a*; only two features, X_r and X_s, are considered, but several intervention types are involved, identified by the plot shapes (i.e. squares, triangles, and circles).

Once the clustering is obtained, in order to determine the most probable type of intervention associated with a triggered alert, the estimated values of its features ($\hat{X}$) are the inputs to the model; the corresponding location P of the asset condition can be inferred; this location will belong to a specific k-cluster based on the distance to its centroid. The information implied in this cluster, which corresponds to the historical interventions of the samples, defines the different maintenance types involved; from a simple rating analysis, a probability for each type can be easily assigned to the alert.

To take into account MMB's know-how, related to the subjective assessment of samples in the k-cluster, the historical occurrence of each maintenance type can be weighted by scores giving more relevance to samples that had subjective evaluations similar to those predicted by the model, with features $[S\hat{X}, C\hat{X}, \hat{G}]$ of the asset implied in the triggered alert, as derived by submodule AM21. Regarding the example, Figure 3*a* shows the k-cluster the sample P belongs to; each sample contained in the cluster carries information regarding historical subjective evaluations $[SX_r, SX_s]$. Those samples with similar subjective evaluation to that predicted will be scored with a higher weight than the rest. In this way, an unsupervised ML approach is complemented with the information provided by the historical know-how. Using this technique, a huge number of assets under similar conditions can be used to plan maintenance strategies, circumventing recurrent problems such as bias.

The methodology described above has serious limitations when data do not show clear grouping, as shown in Figure 3*a*; this drawback can be circumvented by using a second phase based on the KNN algorithm. In this case, the k nearest historical samples to a given triggered alert (e.g. the samples inside the circle in Figure 3*b*) are used define the most probable types of maintenance associated with the alert. The methodology is quasi-similar to the previous one, only with consideration to k samples. This means that the selected samples, including P, belong to the same unique "cluster" in order to apply the technique described in phase one. Following the same approach, the underlined know-how can be considered by giving more importance to samples with a similar evaluation as predicted for sample P, $[S\hat{X}, C\hat{X}, \hat{G}]$, provided by submodule AM21. In the example, each occurrence (from the k-nearest set) is weighted by a score which depends on its subjective assessment; higher for samples with a similar evaluation to P. This methodology is less sensitive to the data distribution than the clustering, but, in general, it uses fewer samples to define the most probable intervention types and some erroneous information can be screened out by altering the value of parameter k.

Fusion Model. To take into account the advantages of the two single methodologies developed—clustering and KNN—for estimating the type of intervention, a merged method has been constructed, summarized in the following steps:

(a) Generate a proposed cluster model, as outlined previously. Figure 3*a* depicts a recreation where clusters are not clearly identified, due to samples close to borders.

(b) Compute the Euclidean distance between sample P and each of the cluster centroids:

$$(d_{P-ce1}, d_{P-ce2}, \ldots, d_{P-ceL}).$$

(c) Choose the k nearest samples to trigger alert P using a criterion distance z between sample P and a generic sample Q belonging to cluster *J-th*, $z_{P-Q} = d_{P-Q} \cdot d_{P-ceJ}$, where d_{P-Q} stands for the Euclidian distance between P and Q. In this way, samples belonging to different clusters are penalized, as it is the case of sample Q in Figure 3*c*.

Figure 3*c* shows the difference between the KNN model (broken circle) and the fusion model (solid closed line). As illustrated, the fusion model is not as dependent on the existence of clear data grouping (i.e. cluster definition) and enables incremental, guided changes to k, regarding samples belonging to the same cluster as P.

As with the previous models, MMB's know-how is taken into account by giving more importance to samples with the same evaluation as the predicted one for sample P, $[S\hat{X}, C\hat{X}, \hat{G}]$ as inferred by submodule AM21.

Empirical Case Study

The pilot case selected is a meshed road network in the central region of Portugal, managed by Infraestruturas de Portugal, totaling 539 km; it includes several road categories, as principal itineraries, supplementary itineraries, national roads, regional roads, and other roads; the road categories are classified on the basis of features such as travel speed, traffic volume, traffic mix, and strategic importance. Regarding traffic levels, the chosen demonstration case presents heterogeneity among the chosen sub-networks (i.e. itineraries), between 2,500 and 10,000 vehicles per day, with an average of 9% heavy vehicles.

Relevant Information and Datasets

The conducted strategy categorized maintenance interventions as either major or routine. Major works are critical in terms of cost, length, and complexity, with medium-term planning (five-year periods) and an annually reviewed prioritization process. Routine maintenance includes smaller scale and lower complexity works, such as pavement repairs, drainage system cleaning, shoulder treatment, minor works performed on bridges, and any urgent repair.

Needless to say, the recorded data underwent a filtering process to extract the relevant information free of inconsistencies (e.g. unidentified/undetermined geometrical location of measurements, measurements captured by different auscultation vehicles, unclear description of the intervention carried out, assets affected), which may reduce the final number of valid records to a non-negligible percentage. The expected outcome is a set of alerts, each linked to a certain infrastructure component, where, in order to keep a certain level of infrastructure condition, maintenance is necessary. The alerts should also indicate the type of maintenance which is needed.

The number of filtered records in the historical interventions database is limited due to the network extension of the pilot case and the time period. In order to circumvent this drawback, a multiplicity of simulated datasets, derived from the statistical distributions of the available empirical data, was generated in order to select the most appropriate ML model.

Model Selection for Estimating Maintenance Interventions

This section presents the results after applying four separate ML techniques: DT, KNN, SVM, and ANN. The main objective is to build models with good generalization capabilities to perform well on new data (i.e. test data for which the model has not been trained). Another objective is to determine the required training-set size for each model to be capable of generalizing. The performance of these classification models has been evaluated by the confusion matrix (*33*).

The relationships between training-set size, model complexity, and prediction error have been analyzed and identified. Two separate goals are addressed:

(a) Model selection: estimating the performance of different models and their parameters, in order to choose the optimal model.
(b) Model assessment: estimating the prediction error of the chosen model when applied to new data.

The dataset, with a sample of 2,000 interventions, was randomly divided into three sets for: training the models, validating prediction errors when choosing a model, and testing the error of the chosen model. In DT techniques, the main parameter to be calibrated is the number of splits or branches, ranging from 2 to 40. The final best model has 28 splits. For the KNN technique, a total of nine models that range from 2 to 20 neighbors have been analyzed. The final calibration of this method includes the choice of additional parameters such as distance metric (Euclidean) and distance (square inverse) weighting function. The SVM technique is mainly characterized by its kernel function (linear, quadratic, three-degree polynomial, gaussian, etc.). Through testing, it is concluded that a three-degree polynomial kernel function is the best option. The ANN technique is a more complex algorithm from the calibration point of view, due to the number of parameters to be configured. Only the number of hidden layers and the number of neurons are addressed here. A pattern neural model, with Levenberg-Marquardt backpropagation training and a mean squared normalized error performance function, was selected. Figure 4*a* shows seven series representing different ANN designs, with variations as follows:

(a) 1 h-layer: only a hidden layer, where the number of neurons range from two to 20.
(b) 2 h-layers [2 X]: two hidden layers, where the first hidden layer has two neurons and the second has a number of neurons ranging from two to 10.
(c) 2 h-layers [3 X]: two hidden layers, where the first hidden layer has three neurons and the second has a number of neurons from two to 10; and so forth.

In order to compare the different models, the complexity parameter is defined as the sum of the weights to be trained in each model. Model complexity increases as the number of neurons or hidden layers increase. Figure 4*a* indicates that the models with only one hidden layer (1 h. layer) achieve better performance; while several models reach high accuracy, the optimal model is one that achieves high accuracy with minimal complexity. The final model chosen has one hidden layer with only six neurons (black diamond sample in Figure 4*a*); the complexity is 78, unlike the model with eight neurons, which reaches a similar accuracy but with complexity of 104.

The final models selected of each technique achieve the following accuracies: DT, 0.989; KNN, 0.960; SVM, 0.969; ANN, 0.969; thus DT is the model with the highest accuracy. The expected test error of each estimated model is calculated using learning curves (Figure 4*b*). These plots illustrate the importance of assessing the ability of a learning technique to generalize. As the training-set sizes get larger, the curves converge toward a threshold representing the amount of irreducible error in the data. The DT model achieves the smallest error. SVM, ANN, and KNN models yielded similar results; none of them reached a test-set average error below 5%. The DT model reached an average test-set error of less than 5% using a training-set size of 500 records. Thus, this model was selected to infer the result of the use case.

In order to understand where the errors in predicted intervention type lie, a confusion matrix was calculated for model DT (Figure 5*a*). This matrix illustrates that,

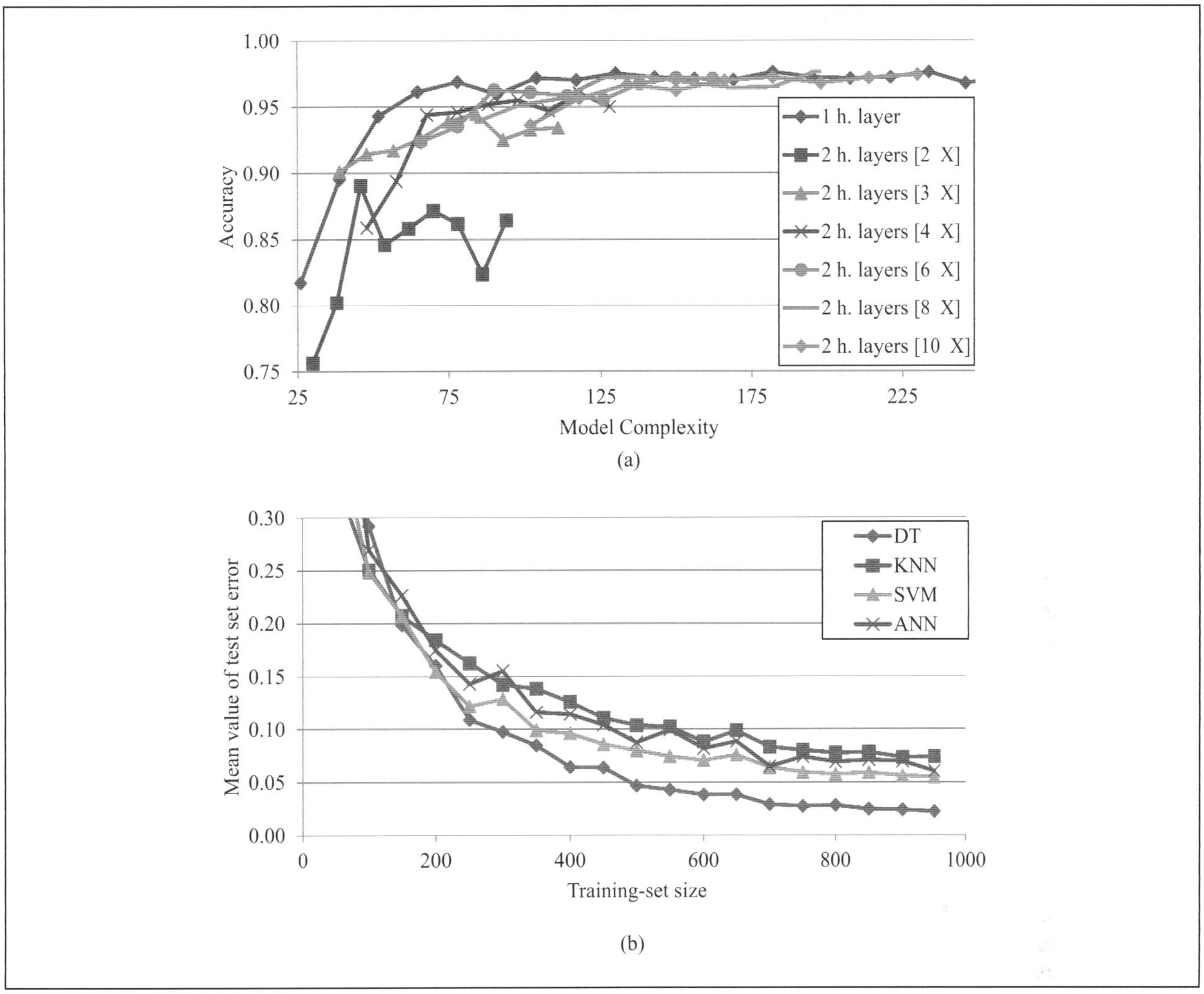

Figure 4. Model selection: (*a*) analysis of ANN method; (*b*) learning curves.

most commonly, errors appear when the model estimates T3 or T4 maintenance types instead of T3.1, and when T2 is predicted instead of T1.

Results Inferred from Empirical Data

In order to check the accuracy of the final selected model, the 2014 campaign dataset is used as a testing sample. Figure 5*b* shows a comparison between the real interventions carried out and the predicted ones. According to Figure 5, the results are reliable, with errors occurring between T4 and T3.1 maintenances.

A total test-set of 1,241 sections (each 500 meters in length) were input to the DT model. In 1,159 of these sections the predicted values coincide with the real ones, obtaining a final accuracy of 93.4%.

Comparing the predicted work orders (WO) and the real ones in all sections: 843 sections out of 859 were correctly predicted as T0 maintenance type; 39/41 as T1; 148/187 as T2; 28/30 as T3; 80/93 as T3.1; and 21/31 as T4. This implies that the most common errors are for model prediction of T2 (only 79.1% of accuracy) and T3.1 (acc. 86%), as was suggested by the confusion matrix (Figure 5*a*).

Conclusions and Prospective Horizons

In this paper, road network maintenance was studied from a predictive intervention approach. Four ML techniques were analyzed to substantiate the optimum choice of predictive model. The automatic learning methodology was trained from a database of WOs, asset features, and measurement auscultations from historical interventions. The main outcomes are:

(a) the estimated intervention type for each road section and the probability of occurrence.

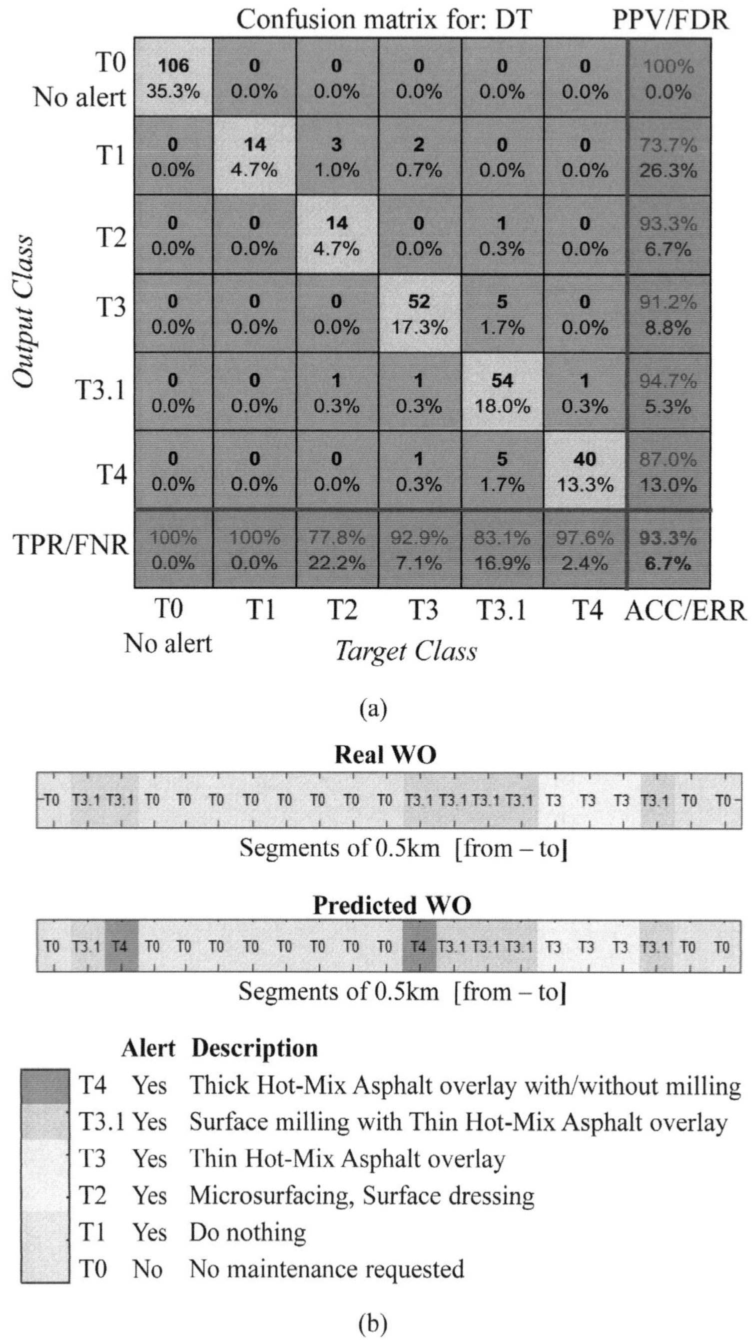

Figure 5. Results of DT model: (*a*) confusion matrix; (*b*) real versus predicted work orders (WO).

(b) a ranked listing of estimated alerts according to the TSL. Each prediction-set is referred to a future scenario identified by its timestamp.

The results evidence that—with a limited historical database, an optimum model choice, parameter fitting, and proper training of the supervised models—the framework methodology provides good predictive capabilities.

The methodologies and results presented herein are far from being exhaustive or conclusive, and several parallel lines of research are open. Among others:

(a) the sensitivity to the quality of intervention description in the historical database, regarding the intervention timestamp.
(b) the importance of the level of detail in the description of the asset condition prior to intervention.
(c) self-learning rules from automatic learning from false positives/negatives.

The methodology presented constitutes a step forward in the generation of a smart decision-making support tool, which can derive intervention plans based on alert forecasting and optimal selection of activities regarding the most critical interventions to be carried out. These are single steps toward the final purpose of covering the full range of planning maintenance at operational, tactical, and strategic levels, under an expert intelligent framework.

Acknowledgments

This research has received funding from European Union's Horizon 2020 Research and Innovation Programme (grant agreement no. 636496). Some of the authors express their gratitude to the Spanish Ministry of Economy and Competitiveness for the partial subsidy granted under the national R&D program (TRA2015-65503) and the Torres Quevedo Programme (PTQ-13-06428). The authors acknowledge Infrastruturas de Portugal (IP) for making available the database used in this research.

Author Contributions

All authors confirm contribution to the paper in the following issues: methodology conception and design, data collection and preparation, ML modeling, analysis and interpretation of results, and draft manuscript preparation. All authors reviewed the results and approved the final version of the manuscript.

References

1. Ahern, A., and G. Anandarajah. Railway Projects Prioritisation for Investment: Application of Goal Programming. *Transport Policy*, Vol. 14, No. 1, 2007, pp. 70–80.
2. Li, M. Y., K. Q. Han, X. Y. Zhang, X. I. Li, Y. J. Zhai, and W. H. Yu. A Management Information System for Mine Railway Transportation Equipment. *Journal of China University of Mining and Technology*, Vol. 18, No. 3, 2008, pp. 373–376.
3. Link, H. An Econometric Analysis of Motorway Renewal Costs in Germany. *Transportation Research Part A: Policy and Practice*, Vol. 40, No. 1, 2006, pp. 19–34.
4. Coffen-Smout, S., and G. J. Herbert. Submarine Cables: A Challenge for Ocean Management. *Marine Policy*, Vol. 24, No. 6, 2000, pp. 441–448.
5. Cagno, E., F. Caron, M. Mancini, and F. Ruggeri. Using AHP in Determining the Prior Distributions on Gas Pipeline Failures in a Robust Bayesian Approach. *Reliability Engineering & System Safety*, Vol. 67, No. 3, 2000, pp. 275–284.
6. Creig-Smith, S. To Refurbish or Replace Steel Water Pipelines, That Is the Question. *Engineering Failure Analysis*, Vol. 8, No. 2, 2001, pp. 107–112.
7. Castanier, B., and M. Rausand. Maintenance Optimization for Subsea Oil Pipelines. *International Journal of Pressure Vessels and Piping*, Vol. 83, No 4, 2006, pp. 236–243.
8. Brito, A. J., and A. T. de Almeida. Multi-Attribute Risk Assessment for Risk Ranking of Natural Gas Pipelines. *Reliability Engineering & System Safety*, Vol. 94, No. 2, 2009, pp. 187–198.
9. IBM Maximo. http://www-03.ibm.com/software/products/en/maximoassetmanagement. Accessed June 6, 2017.
10. Infor Corporation. USA. http://www.infor.com/solutions/eam/. Accessed November 11, 2017.
11. Optram. *Bentley Systems Inc.* USA. https://www.bentley.com/en/capabilities/apm/analyze-rail-track. Accessed November 11, 2017.
12. Oracle, EAM. https://docs.oracle.com/cd/E18727_01/doc.121/e13670/T259967T259970.htm. Accessed November 11, 2017.
13. SAP. https://www.sap.com/products/linear-asset-management.html. Accessed November 11, 2017.
14. Kuennen, T. *Pavement Preservation Compendium II.* FHWA-IF-06-049. FHWA, U.S. Department of Transportation, Washington, D.C., 2006.
15. FHWA. *Life-Cycle Cost Analysis in Pavement Design (In Search of Better Investment Decisions)*. FHWA-SA-98-079. FHWA, U.S. Department of Transportation, Washington, D.C., 1998.
16. SUST. *Sustainable Bridges - Assessment for Future Traffic Demands and Longer Lives.* FP6 collaborative project, 2003-2007. European 6th RTD Framework Programme. https://trimis.ec.europa.eu/project/assessment-future-traffic-demands-and-longer-lives. Accessed November 11, 2017.
17. Shapiro, S. *Foundation without Foundationalism. A Case for Second-Order Logic.* Clarendon Press, Oxford University Press, Oxford, UK, 1991.
18. Peter, J. *Introduction to Expert Systems*, 3rd ed. Addison Wesley Longman, Harlow, UK, 1999.
19. Negnevitsky, M. *Artificial Intelligence: A Guide to Intelligent Systems*, 2nd ed. Addison-Wesley, London, UK, 2005.
20. Yella, S., M. S. Dougherty, and N. K. Gupta. Artificial Intelligence Techniques for the Automatic Interpretation of Data from Non-Destructive Testing. *Insight – Non-Destructive Testing and Condition Monitoring*, Vol. 48, No. 1, 2006, pp. 10–20.
21. Podder, T. *Analysis & Study of AI Techniques for Automatic Condition Monitoring of Railway Track Infrastructure.* Master thesis. Högskolan Dalarna, Sweden, 2010.
22. Iqbal, Z. *A Study of AI Techniques for Railheads, Vegetation, Switches & Crossings.* Master thesis. Högskolan Dalarna, Sweden, 2010.
23. Sambath, S., P. Nagaraj, and N. Selvakumar. Automatic Defect Classification in Ultrasonic NDT Using Artificial Intelligence. *Journal of Nondestructive Evaluation*, Vol. 30, No. 1, 2011, pp. 20–28.

24. Martín, O., M. López, and F. Martín. Artificial Neural Networks for Quality Control by Ultrasonic Testing in Resistance Spot Welding. *Journal of Materials Processing Technology*. Vol. 183, Vol. 2–3, 2007, pp. 226–233.
25. Kirkman, M. M. *Maintenance of Railways*. C.N. Trivess Printers, Chicago, IL, 1886.
26. Morrow, L. C. *Maintenance Engineering Handbook*, 1st ed. McGraw-Hill, New York, NY, 1957.
27. Witten, I. H., E. Frank, and M. A. Hall. *Data Mining: Practical Machine Learning Tools and Techniques*. Morgan Kaufmann Publishers, San Francisco, CA, 2011.
28. Dick, C. T., C. P. L. Barkan, E. R. Chapman, and M. P. Stehly. Multivariate Statistical Model for Predicting Occurrence and Location of Broken Rails. *Transportation Research Record: Journal of the Transportation Research Board*, 2003. 1825: 48–55.
29. Podofillini, L., E. Zio, and J. Vatn. Risk-Informed Optimisation of Railway Tracks Inspection and Maintenance Procedures. *Reliability Engineering and System Safety*, Vol. 91, No. 1, 2006, pp. 20–35.
30. Adarkwa, O. A., and N. Attoh-Okine. Pavement Crack Classification Based on Tensor Factorization. *Construction and Building Materials*, Vol. 48, 2013, pp. 853–857.
31. Faiz, R. B., and E. A. Edirisinghe. Decision Making for Predictive Maintenance in Asset Information Management. *Interdisciplinary Journal of Information, Knowledge, and Management Volume*, Vol. 4, 2009, pp. 23–36.
32. SAS. *Predictive Asset Maintenance*. SAS Institute Inc. World Headquarters. https://www.sas.com/es_es/search.html?q=Predictive%20Asset%20Maintenance. Accessed November 11, 2017.
33. Stehman, S. V. Selecting and Interpreting Measures of Thematic Classification Accuracy. *Remote Sensing of Environment*, Vol. 62, No. 1, 1997, pp. 77–89.

The Standing Committee on Transportation Asset Management (ABC40) peer-reviewed this paper (18-04092).

The content of this paper reflects the authors' views alone and it is hereby stated that neither the EU nor Infrastruturas de Portugal is liable for any use that may be made of the information contained in it.

Article

TRR
JOURNAL OF THE TRANSPORTATION RESEARCH BOARD

Transportation Research Record
2018, Vol. 2672(44) 55–61

DOI: 10.1177/0361198118782764
journals.sagepub.com/home/trr
SAGE

Risk Management and Data Needs: A State of the Practice Survey of State Highway Agencies

Inya Nlenanya[1] **and Omar Smadi**[1]

Abstract

Risk management analysis is one of the new requirements under MAP-21 that separates transportation asset management programs from business as usual for the state departments of transportation (DOTs). Based on this requirement, each agency will discuss the concept of risk and how it should be incorporated into its transportation asset management program as well as how it informs maintenance practices, asset replacement or rehabilitation, and emergency management and response planning. This will require an agency to provide a list of risk exposures and document its system-wide risk management strategy. This paper presents the results of a state of the practice survey of how agencies are developing their risk-based asset management plan and discusses recommendations for future research. The results show that state highway agencies are increasingly adapting the way they do business to include explicit considerations of risks. At the moment, this consideration of risk is not linked to data, and as a result most agencies do not have a data driven way of tracking risk and updating their risk exposures. The significance of the results highlights the need for further research on data driven risk management and to synthesize methodologies for integrating risk assessment into an agency's decision-making process.

On July 6, 2012, the Moving Ahead for Progress in the 21st Century Act (MAP-21) was signed into law by President Obama. MAP-21 and the Fixing America's Surface Transportation Act (FAST) which supersedes MAP-21, require states to develop and implement a risk-based asset management plan for the national highway system (NHS) to improve or preserve the condition of the assets and the performance of the system. The legislation focuses on the development of a transportation asset management program (TAMP) for bridges and pavements on the NHS, as well as other infrastructure assets within the right-of-way corridor (*1*, *2*).

Risk management analysis is one of the new requirements under MAP-21/FAST that separates TAMP from business as usual for the state departments of transportation (DOTs). As a result of this requirement, each agency will discuss the concept of risk and how it has been incorporated into its TAMP as well as how it informs maintenance practices, asset replacement or rehabilitation, and emergency management and response planning. This will require an agency to provide a list of risk exposures and document its system-wide risk management strategy. The intent of this paper is to understand how each state is currently integrating risk management into their TAMP, what available data informed the risk identification process and what additional data is needed to properly execute the risk assessment process. To accomplish this, an online survey of state DOTs was conducted to determine how risk management is being integrated. In this paper, we summarize the responses to the survey as well as discuss its implications for risk-based asset management.

Background

Risk is defined as the positive or negative effects of uncertainty upon agency objectives (*3*). Risk is anything that minimizes one's ability to achieve stated goals and objectives. Although the entire concept of risk management is to control potential negative outcomes, in the case of positive risks, it occasionally deals in opportunity (*4*) by supporting activities that foster innovation or bringing about the greatest returns on investment (*5*). Hence, risk management is a process of analytical and management activities that focus on identifying and responding to the inherent uncertainties of managing a complex organization and managing capital facilities (*3*, *6*). In the context of transportation agencies, it generally consists of the strategic and systematic process of operating, maintaining, and improving physical assets and managing their highway network with an emphasis on minimizing threats and maximizing opportunities. Hence, a risk-based asset management approach is tasked with identification of risks

[1]Institute for Transportation, Iowa State University, Ames, IA

Corresponding Author:
Address correspondence to Inya Nlenanya: inya@iastate.edu

that affect the DOT's ability to meet its stated goals, assessment of the impacts of the risks, as well as proffering mitigation (*4*). At the core of this is the ability for transportation agencies to have a mechanism for measuring the identified risks, as this will help drive the possible mitigation activities.

These risks are generally a result of anticipated future outcome dependent on metrics of both present conditions as well as future predictions. They cut across technology, human resources, weather, and environmental conditions among other things. Colorado DOT considered, as part of their TAMP process, the following three levels of risk (*7*):

1. Agency (strategic, corporate) risks—affects mission, vision, and overall results of the asset management program. Examples include politics, public perception, reputation, levels of available revenue, etc.
2. Programmatic (business line) risks—affects DOTs' ability to deliver projects and meet targets within a program. These may include organizational and systemic issues as well as revenue and economic uncertainties that in general cause project delay. These issues can be multivariate. Examples include project delivery risks, revenue uncertainties, cost-estimating processes, revenue and inflation projection inaccuracies, construction cost variations, materials price volatility, data quality, retirements, etc.
3. Project/asset risks—affects scope, cost, schedule, and quality of projects. In contrast to programmatic risks, project risks are related to specific projects. In other words, there are inherent issues in a given project that may result in a project delay. Examples include hazardous materials, geology, environmental issues, right-of-way issues, utilities, project development timeline/delays, scope growth, cost overruns, project delays, etc.

Although all the above levels of risk are clearly defined, there exists significant overlap in what an agency needs to do at every level. The identification of the actual data needed in each level for risk identification and analysis is a very challenging task. This is partly due to the substantial overlap between the various levels, as well as the lack of relevant research initiatives in this field (*8*).

Nevertheless, an astute understanding of these levels of risk is a very good starting point for risk management. The ISO Risk Management Framework (*6*) (Figure 1), is very helpful to breaking down the broad categories of risk into actionable pieces for the agency to tackle.

They include:

1. Establish Context: This means understanding what aspect the risk covers. It could be social, environmental, economic, legal, cultural, etc. This also identifies the specific policy goals or objectives that the risk undermines.

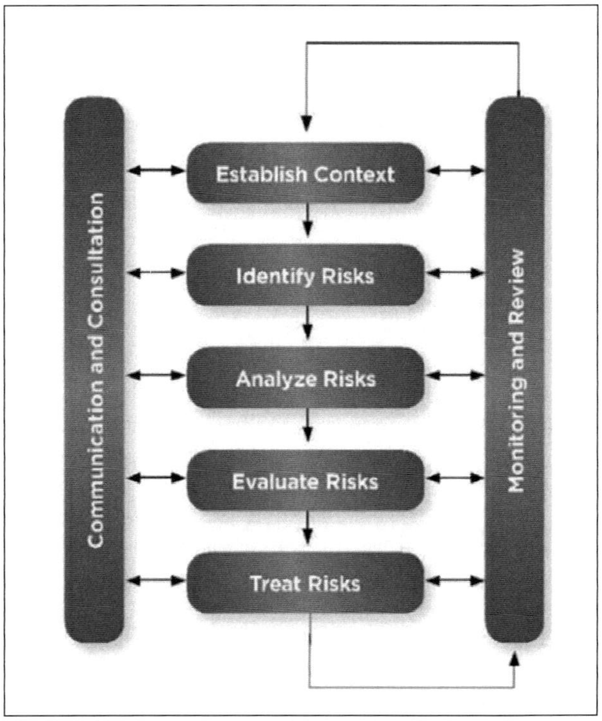

Figure 1. ISO risk management framework.

2. Identify Risks: Identify the risks to the assets with respect to the agency's management goals. This process discloses and documents all foreseeable risks and their triggers that could affect policy goals and objectives.
3. Analyze Risks: Find out what impact this risk will have on the overall asset management plan. This process determines the likelihood and consequence of failure.
4. Evaluate Risks: This stage helps to firm up decisions taken by comparing magnitude of risk with its risk tolerance based on its relative importance to the agency, program, or project.
5. Mitigate Risks: This decision-making step applies the "five Ts" which are to treat, tolerate, terminate, transfer or take advantage of the risk. This involves determining appropriate mitigation strategies as well as monitoring strategies, and re-evaluating risks.

Effective risk management is only possible if the appropriate methods, which address the root cause of the identified risk, are utilized in its mitigation. All agencies have a common goal of achieving a stated level of low risk to their assets and this will require constant communication and consultation with the affected stakeholders as well as constant monitoring and review by the agency. To this end, this research provides insights into how the agencies are integrating risk into their TAMP and provides actionable intelligence on how to further refine the process.

Research Objectives

The primary objectives of this research were to investigate the state of the practice for integrating risk management in U.S. highway agencies' TAMP and to make recommendations for future research into bridging the resulting gaps. Findings address the TAMP readiness of the agencies, the risk identification process, and what available data were used to generate the risk registers. Objectives of this research also included the following:

- Complement the research conducted for the NCHRP 08-93: Managing Risk across the Enterprise—A Guidebook for State Departments of Transportation (*9*) as well as all prior studies in this area
- Determine research needs to assist state DOT compliance with MAP-21 and FAST Act requirements

Results

An online survey was prepared and sent to all 50 state highway agencies to understand how each state is currently integrating risk management into their TAMP, what available data (hard data or soft data) informed the risk identification process, and what additional data is needed to properly execute the risk assessment process. The unit of the survey was the state DOT. The scope of the study was limited to the state DOTs. The survey covered three main topics: TAMP readiness, risk identification methodology, and linking risk identification and assessment to data. The questions were prepared by using various formats, such as radio buttons, check-all-that-apply boxes, ranking, and short essay question fields. The ranking questions employed randomization of the options to eliminate bias due to order. Some questions asked recipients to elaborate on their selections in essay fields or to provide supplementary information. Furthermore, the recipients were asked to provide their contact information if they were willing to permit the survey team to follow up on their responses.

A total of 26 completed surveys from 50 States were received. Therefore, the response rate was 52%.

TAMP Readiness

This section addresses how far along the agencies are in terms of developing their risk-based TAMP framework.

Asset Management Task Force. An asset management group or task force is usually the first step towards formalizing an agency's commitment to asset management and institutionalizing transportation asset management (TAM) as a business process. More than 90% of the responders confirmed the existence of a task force to coordinate all of their asset management activities. The high percentage is an improvement compared to a previous study published in 2013 (*10*). A follow-up question on the composition of the task force was asked and the responses were consistent with what was reported in 2013 (*10*). The only difference is that most agencies are reorganizing and refocusing the asset management group to not only have a policy-oriented function but also include a steering committee or a technical working group tasked with implementation. The policy side is made up of division heads/administrators and unit leaders that encompass the following disciplines: bridge, pavement, planning and programming, operations, research, IT, design, materials, data management, and finance. They are responsible for developing the TAMP.

On the other hand, the steering committee or technical working groups are made up of the data owners, data managers, and an asset liaison from each of the agency's asset class to serve as a line of communication between the asset and the technical working group to help facilitate proactive asset management. The steering committee is responsible for the implementation of the TAMP. This can be evidence that more and more agencies are taking ownership of the TAMP process and reorganizing it in the best way that fits with their unique situations.

Risk Management Task Force. With the requirement under MAP-21/FAST for risk-based asset management, 70% of DOT responders have the asset management group taking the lead on the development and implementation of the risk management portion of the TAMP. This is the case since the many of the identified risks and strategies are foundational for the TAMP development. A follow-up question on the composition of the risk management group was asked to see if agencies were incorporating the risk requirements as part of the overall TAMP or as an independent part. The responses showed that the reorganization of the asset management group was due to the accommodation of the risk requirements. The main reason for the technical working group was to create a more focused group working on aligning risk management with the agency's overall goals. Regardless of whether the asset management group is in charge of the risk management activities or not, most responders agree that there should be a clear line of communication between the risk management activities and the overall agency policy and goals to make sure that strategies and initiatives are in sync with responsibilities.

Linking Strategic Goals to Risk. From the discussion on TAMP readiness, there is obvious momentum among state highway agencies to implementing risk-based asset management. However, there is need to understand how a risk is assigned to either agency, program, or project level. Is it based on its scope or magnitude of its impact? This is important because risk management concept requires that risk be treated at its highest level (*11*). It is likely that a given risk can impact an agency at both the program and agency level. In that case, the risk should be mitigated at the agency level.

In a follow-up question, one agency responded that risk identification was conducted by asset category and then sorted into the various levels of the risk management program. This undoubtedly allows the agency to upscale or downscale a risk depending on their understanding of its scope and magnitude.

In addition, it is imperative to link strategic goals to risks to ensure that risk is identified and placed at its highest level. However, 55% of the responders linked strategic goals to risk, just two agencies more than those that treated it independently. The small gap can be explained by the fact that some agencies were actually doing both. This was confirmed in one follow-up interview.

Risk Management Practice

This section focuses on how state highway agencies are integrating risk into their TAMP. The application of risk-based asset management as required by MAP-21/FAST is still in its infancy (*12*, *13*). Risk management is the core principle of asset management (*6*). The product of risk management is efficiently using available resources to manage programs through improved communication. Applying risk management to look at decision-making about program delivery makes it possible to identify risk triggers, assess and prioritize those risks, and determine strategies that provide a framework on how best to deal with future issues affecting agency goals (*14*).

When asked about the methodology for the risk identification process, all the responders had a similar methodology for risk identification. Through consultation with senior management, asset managers, subject matter experts, key staff from various asset classes, and in one case an outside consultant, most agencies performed strategic risk identification in workshops. From this risk identification workshop, risk registers were compiled. This confirms the findings in (*3*) that risk register is the most commonly used tool for risk management for all levels of risk. All the draft or final TAMP documents of state DOTs on the AASHTO TAM portal utilized a risk register to execute their risk management.

The risk register was evaluated in terms of the likelihood of a risk event and impact. The scale (Figure 2) shows the risks identified at the workshop with their determination of the likelihood of the event occurring, where 5 is high and 1 is low. The risk was then assessed in terms of its consequence, where 5 is catastrophic and 1 is negligible.

Risk registers and their outputs such as heat maps allow definition of priority risks through risk evaluation, establishing risk tolerances and decision criteria, and using these factors to inform response/treatment. More than 90% of the respondents had developed risk registers that encompass both agency and program-level risks. This is consistent with what was reported in O'Har et al. (*15*). Only 15% of the responders had generated risk registers for all three risk levels.

		Consequence (Level/Descriptor)				
Likelihood		1	2	3	4	5
Level	Descriptor	Negligible	Minor	Major	Critical	Catastrophic
1	Low	1	2	3	4	5
2	Medium Low	2	4	6	8	10
3	Medium	3	6	9	12	15
4	Medium High	4	8	12	16	20
5	High*	5	10	15	20	25

Figure 2. Risk rating scale (*8*).

These risk registers are intended by the agencies to be a living document that is updated as new risks are encountered. Figure 3 summarizes the frequency of risk registers review and update. Across all survey responders, most agencies did not have a set schedule for updating the risk registers while 23% revise annually. Some agencies preferred to review and update as needed. This provides a more complete and updated view of state DOTs' schedule for updating the risk register with 24 agencies responding, than what was reported in O'Har et al. (*15*) with only 13 state DOTs responding. Both agree that most agencies are still unsure how frequently to revise the risk registers.

Risk Identification and Data

This section's purpose is to understand the data that were used to drive the risk assessment process and the additional data that is needed. The monitoring and updating of the risk management process is critical to ensure that new risks are identified and existing risks are tracked and updated (*3*). To help agencies understand the impacts associated with the risks as they occur, agencies need to properly track asset failures. This ability to track and quantify risk would benefit the risk classification process (*16*) and its seamless integration into TAMP.

The agencies were asked about what data (hard data or subjective data) were considered in their risk assessment exercise and to rank its importance. The results in Tables 1–3 for the strategic, program, and project level risks respectively, show only the factors that were consistently ranked in the top three as presented in columns 1, 2, and 3. For strategic risks, the top three driving factors for the risk assessment were historic funding levels, policy changes and political climate, and the financial market (Table 1). Also ranking high was upper management personnel changes. These are consistent with what would impact the success of a TAM implementation (*17*).

Program level risk identification and assessment are driven by project development timelines, federal mandates, and revenue and inflation projections (Table 2). For project level

Table 1. Strategic Risk Data Ranking

Data	1	2	3	Total*
Funding levels (historic)	65%	25%	10%	100%
Policy changes and political climate	15%	15%	25%	55%
Economy—financial market	0%	5%	40%	45%
Legislative requirements	0%	30%	5%	35%
Personnel changes—upper management	5%	15%	10%	30%
Demand/socioeconomic data	15%	5%	5%	25%
Technology changes/limitations	0%	5%	5%	10%
Outsourcing to consultants/vendors	0%	0%	0%	0%

*20 agencies responded.

Table 2. Program Risk Data Ranking

Data	1	2	3	Total*
Project development timelines	7%	27%	20%	54%
Revenue and inflation projections	27%	13%	7%	47%
Federal mandates and legislative expectations	7%	7%	33%	47%
Historic data	13%	0%	20%	33%
Personnel—expertise and experience	13%	13%	7%	33%
Project delivery schedules	13%	7%	7%	27%
Cost of materials	7%	13%	7%	27%
Program management record	13%	13%	0%	26%
Climate change—flooding, temperature change, etc.	0%	7%	0%	7%

*15 agencies responded.

Table 3. Project Risk Data Ranking

Data	1	2	3	Total*
Structural condition	27%	13%	13%	53%
Life cycle costs	13%	7%	33%	53%
Demand/usage	0%	33%	7%	40%
Overall performance	27%	13%	0%	40%
Funding impacts/tradeoffs	13%	7%	20%	40%
Initial agency costs	0%	13%	20%	33%
Functional condition	13%	7%	0%	20%
Location/environmental	0%	7%	7%	14%
External risks—litigation	7%	0%	0%	7%
Customer feedback/complaints	0%	0%	0%	0%
Attributes/characteristics	0%	0%	0%	0%

*15 agencies responded.

risks, the factors consistently ranked in the top three were structural condition, life cycle costs, demand/usage, and overall performance (Table 3). It is expected that physical condition, demand, and performance will dominate project level risk consideration. The overall performance, on the other hand, is a function of condition which is impacted by demand, and how much money is invested. In other words, at the project level, risk assessment is driven by whatever impacts performance. The high ranking of demand/usage in project risk analysis is consistent with its role in project selection decision-making as found in Flintsch and Bryant (*8*).

At all three risk levels, funding and policy-related issues stand out. Agencies are always concerned about available funding to maintain performance levels and still be compliant on the federal mandates or state legislations.

Risk and Resiliency Considerations. When asked if additional data were considered, 27% of responding agencies answered in the affirmative. A list of additional data considered were seismic risk, availability and reliability of performance data, continued use of a systematic assessment approach, and safety data. In addition, two thirds of the agencies considered past risk events and the response in their risk assessment process. Most of these past events were weather-related events. This is evidence that most agencies see value to risk and resiliency considerations in risk management. The combination of risk and resiliency in the risk management process is explored in Flannery et al. (*18*) and Herrera et al. (*19*) and has been beneficial to prioritizing mitigation measures for transportation assets.

Data Needs and Risk Tracking. The majority of agencies, 90%, did not identify data needed for the risk identified. In addition, 60% of the responding agencies did not specify how frequently to track the identified risks in their risk registers (Figure 3), including the three agencies that had a data driven approach to tracking risk (Figure 4). This can be an indication of a lack of understanding of the benefit of a clear and regular update of the risk assessment process.

Conclusions and Research Needs

The goal of this research was to investigate the state of the practice of how agencies are integrating risk in their TAMP as well as identify research needs to facilitate state DOTs' compliance with the MAP-21 and FAST Act. From the state of the practice survey, it is clear that state highway agencies are taking more and more ownership of their TAMP. This is validated in how some agencies are reorganizing their asset management group to include a policy-focused group in charge of development and a technical group in charge of the risk management process and implementation.

Currently, the risk assessment process is driven more by subjective data than by data that is currently tracked. Hence most agencies are unsure on how to track risks. This warrants future research into the synthesis of a composite measure of risk and performance that reflects organizational needs. This composite measure should provide a systematic process for updating the risk assessment process and the risk registers that are generated. It will serve as a basis for comparing alternative improvement strategies (investment and policy approaches) and for tracking results over time.

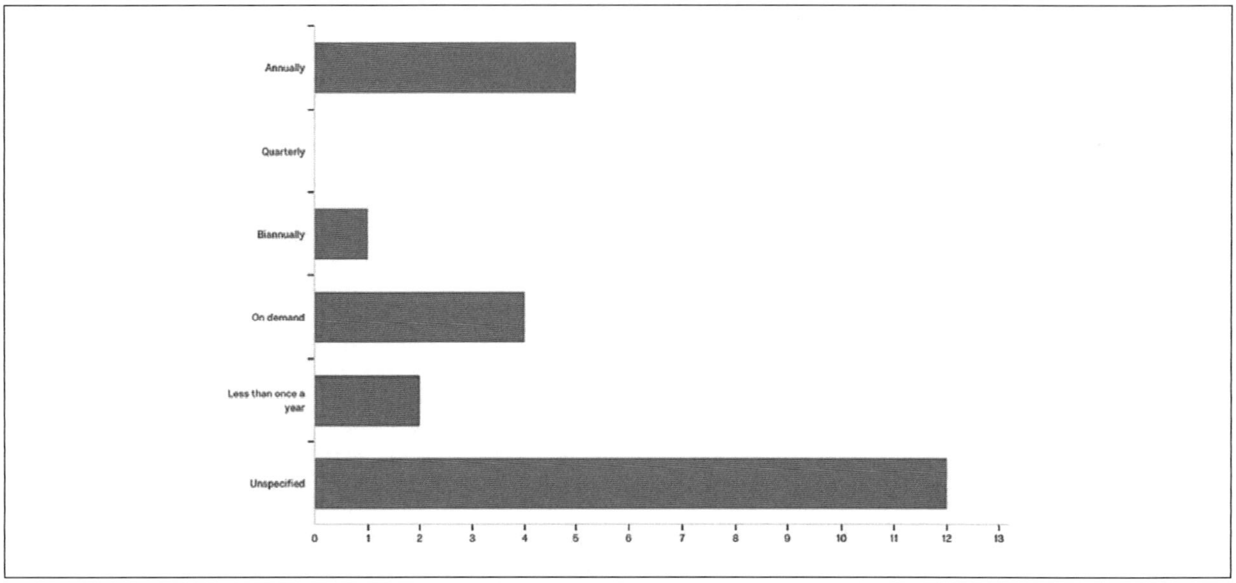

Figure 3. Number of respondents showing frequency of risk register update.

In addition, there is need to link historic data to risk events to produce a rich database that can allow for predictive and prescriptive modelling. Within the context of risk-based TAMP, leveraging available data and analysis tools helps visualize and articulate, in both qualitative and quantitative terms, how the combination of various risks and strategies would influence performance targets. It paves the way for the consideration of how various risks, such as revenue constraints, policy or upper management personnel changes can impact a state highway agency's ability to provide services to its stakeholders.

Agencies' reliance on past events prompts the question, what is the overlap between risk management and resiliency planning? Resilience, in this context, is the ability of a system to recover from an unexpected risk event. How does resiliency planning impact risk? Do the two activities augment each other? At which point is that the case and at what threshold is that no longer the case? The preceding questions underscore the need for understanding the relationship between resiliency and risk considerations in effective risk management. There is need for research to document the extent to which implementation of measures across the risk management framework genuinely helps develop the attributes of a resilient transportation infrastructure.

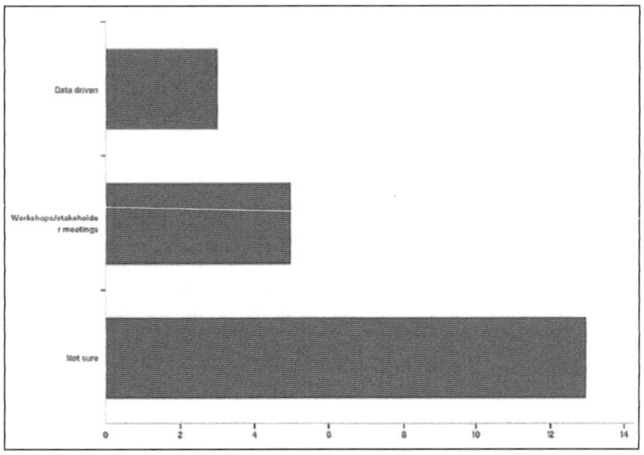

Figure 4. Number of respondents on how to track identified risks.

All agencies are using risk registers as their risk management tool of choice. As a result, it is essential to identify how agencies are using, or intend to use, the risk registers for decision-making. In the process, synthesize methodologies for integrating risk registers into the agency's decision-making process. The overall goal of this endeavor is to minimize the duplication of resources for risk management and to foster a sustainable approach to dealing with uncertainties without undermining organizational efficiency.

Author Contributions

The authors confirm contribution to the paper as follows: study conception and design: Inya Nlenanya, Omar Smadi; data collection: Inya Nlenanya, Omar Smadi; analysis and interpretation of results: Inya Nlenanya; draft manuscript preparation: Inya Nlenanya, Omar Smadi. All authors reviewed the results and approved the final version of the manuscript.

References

1. Congress, US. *Moving Ahead for Progress in the 21st Century (MAP-21)*, 2012, pp. 113–176, 1201, 1202.
2. AASHTO. *Summary of the New Surface Transportation Bill: Fixing America's Surface Transportation (FAST) Act*. AASHTO, Washington, D.C., 2015.

3. D'Ignazio, J., M. Hallowell, and K. Molenaar. *NCHRP Project 20-24(74): Executive Strategies for Risk Management by State Departments of Transportation*. Transportation Research Board of the National Academies, Washington, D.C., 2011.
4. Proctor, G. D., and S. Varma. *Risk-Based Transportation Asset Management Literature Review*. FHWA-HIF-12-036, 2012.
5. Berg, H. P. Risk Management: Procedures, Methods and Experiences. *Risk Management*, Vol. 1, No. 17, 2010, pp. 79–95.
6. International Organization for Standardization (ISO). *ISO 31000 Risk Management: Principles and Guidelines*. International Organization for Standardization (ISO), Geneva, Switzerland, 2009.
7. *Colorado Department of Transportation Risk-Based Asset Management Plan 2013*. https://www.codot.gov/programs/colorado-transportation-matters/documents/risk-based-transportation-asset-management-plan.pdf. Accessed November 9, 2017.
8. Flintsch, G. W., and J. W. Bryant. *Asset Management Data Collection for Supporting Decision Processes*. US Department of Transport, Federal Highway Administration, Washington, D.C., 2006.
9. Transportation Research Board. *NCHRP Report 08-93: Managing Risk Across the Enterprise: A Guidebook for State Departments of Transportation Final Literature Review*. Transportation Research Board, Washington, D.C., 2016.
10. Hawkins, N. R., and O. Smadi. *NCHRP Synthesis 439: Use of Transportation Asset Management Principles in State Highway Agencies*. Transportation Research Board of the National Academies, Washington, D.C., 2013.
11. Curtis, J., D. D'Angelo, M. Hallowell, T. Henkel, and K. R. Molenaar. Enterprise Risk Management for Transportation Agencies. *Transportation Research Record: Journal of the Transportation Research Board*, 2012. 2271: 57–65.
12. Federal Highway Administration (FHWA). *Risk-Based Transportation Asset Management: Evaluating Threats, Capitalizing on Opportunities*. FHWA-HIF-12-035, U.S. Department of Transportation, Washington, D.C., 2012.
13. McGuire, T. B. *Enterprise Risk Management in State Departments of Transportation*. Civil Engineering Graduate thesis and dissertations. Paper 7, University of Colorado, Boulder, CO, 2014.
14. Federal Highway Administration (FHWA). *Asset Management Overview*. FHWA-IF-08-008, Washington, D.C., 2007.
15. O'Har, J. P., C. W. Senesi, and K. R. Molenaar. Development of a Risk Register Spreadsheet Tool for Enterprise-and Program-Level Risk Management. *Transportation Research Record: Journal of the Transportation Research Board*, 2017. 2604: 19–27.
16. Boadi, R. S. *Toward a Risk Framework for Prioritizing Ancillary Transportation Assets for Management*. Doctoral dissertation. Georgia Institute of Technology, Atlanta, 2011.
17. Akofio-Sowah, M., and A. Amekudzi-Kennedy. Identifying Factors to Improve Transportation Asset Management Program Sustainment: Applying Implementation Research and Change Management Principles. *Transportation Research Record: Journal of the Transportation Research Board*, 2016. 2593: 1–7.
18. Flannery, A., M. Pena, M. Katon, and J. Kennedy. Risk and Resilience Analysis for Emergency Projects. *Transportation Research Record: Journal of the Transportation Research Board*, 2015. 2532: 74–82.
19. Herrera, E., A. Flannery, and M. Krimmer. Risk and Resilience Analysis for Highway Assets. *Transportation Research Record: Journal of the Transportation Research Board*, 2017. 2604: 1–8.

The Standing Committee on Transportation Asset Management (ABC40) peer-reviewed this paper (18-05169).

Article

TRR
JOURNAL OF THE TRANSPORTATION RESEARCH BOARD

Transportation Research Record
2018, Vol. 2672(44) 62–71
© National Academy of Sciences: Transportation Research Board 2018
Article reuse guidelines:
sagepub.com/journals-permissions
DOI: 10.1177/0361198118787088
journals.sagepub.com/home/trr
SAGE

Framework for Interactive M^3 Visualization of Microscopic Traffic Simulation

Hong Yang[1], Yuzhong Shen[2], Mahmud Hasan[3], Daniel Perez[3], and John Shull[3]

Abstract

Full comprehension of microscopic traffic simulation (MTS) models has necessitated the development of proper visualizations. Existing MTS models only provide limited capability of two- and/or three-dimensional displays that often restrict users' viewpoint to a flat screen. Their downscaled scenes neither provide a realistic representation of the environment nor allow different users to simultaneously experience the simulation model from different perspectives. This largely prevents analysts from effectively demonstrating and disseminating their simulation results to various stakeholders of different background and knowledge. In light of these issues, this paper aims to develop a framework that enables a multi-user, multi-perspective, and multi-mode (M^3) visualization architecture for microscopic traffic simulation. The proposed framework is empowered by the latest advances in cloud computing and virtual reality (VR) to support interactive and immersive visualization for simulated traffic environments. A client-server architecture allows multiple users at distributed physical locations to view the same simulation from multiple perspectives simultaneously and supports a variety of virtual/augmented reality devices. A prototype of the proposed M^3 visualization framework is implemented and demonstrated by simulating and visualizing a model of typical traffic operations in a high-density urban area. The promising capability of the M^3 visualization framework is attested. Potential improvements over the present study to further excel current visualization framework are also discussed.

The continued evolution of transportation systems has necessitated the extensive use of microscopic traffic simulation (MTS) models worldwide. The level of detail and accuracy required to address stakeholders' questions is often beyond the capabilities of many analytical models. Meanwhile, MTS models are warranted because of their high accuracy and fidelity to actual traffic conditions. Their versatility has been attested in many successful applications for addressing from traditional operational problems (*1*) to the latest connected and autonomous vehicle (CAV) technologies (*2*, *3*). In all applications, engineers often need to convey meaningful findings to various stakeholders. In practice, much effort has been devoted to statistical analysis of simulation output data. This is crucial in quantitatively comparing different scenarios for decision making (*4*). The success of the communication heavily relies on the ability of the engineers to interpret and explain the stories behind the various statistics. However, it is often challenging for stakeholders with different backgrounds (e.g., administrators, planners, citizens, etc.) to easily perceive the stories and anticipate issues that may occur in each modeled scenario. When considering such an issue, visualization of simulated scenarios is increasingly used to provide a common communication medium to facilitate deep understanding that crosses knowledge boundaries between engaged stakeholders.

In fact, the importance of simulation visualization in decision makings and public outreach has been evidenced in many projects. According to a recent study that examined the best practices in the use of MTS models, most surveyed transportation agencies (87%) used visual animations as a means to present simulation results to stakeholders (*5*). More than half of them believed that visualization is more effective at conveying simulation results than conventional charts or maps. Users reached a consensus that the main power of simulation visualization lies in that it enables users to review the animations and perceive any network view to confirm their observations of traffic conditions (*5*, *6*). This greatly helps initiate dialogues among different groups of stakeholders.

Most MTS models are now built with graphical user interfaces (GUIs) that provide displays of the simulation models

[1]Department of Modeling, Simulation & Visualization Engineering, Transportation Research Institute, Old Dominion University, Norfolk, VA
[2]Department of Modeling, Simulation & Visualization Engineering, Department of Electrical and Computer Engineering, Old Dominion University, Norfolk, VA
[3]Department of Modeling, Simulation & Visualization Engineering, Old Dominion University, Norfolk, VA

Corresponding Author:
Address correspondence to Yuzhong Shen: yshen@odu.edu

in either 2-dimensional (2D) or 3-dimensional (3D) environments. These display modules, however, often confine users to viewing simulation via traditional 2D interfaces such as computer screens. Even with larger display devices (e.g., projector screens), the rendered simulation models have to be scaled down. The compressed scenes unavoidably lead to visually perceivable distortions. Moreover, although many simulation tools provide possible interactions through a keyboard and/or mouse, the options are often limited to adjusting the single-person viewpoint, zooming levels, and so forth. This restricts users from freely making real-time adjustments to the simulation scenarios. In addition, the built-in GUIs lack the ability to place users in an immersive 3D environment. Alternatively, animated video clips are often recorded by a camera navigating along predefined paths in the simulation scene. This imposes another problem that only a fixed number of views can be demonstrated. The audience can only view the same scene when playing the video. Furthermore, existing traffic visualization can only be run on the same computer on which the traffic simulation is executed, severely limiting its capabilities to present the results to audiences at different geographical locations.

Lately, the continual advancements in computer graphics, visualization, simulation, and computing hardware have made it easier to emulate complex 3D environments in real time. Thus, the potential to directly link MTS to a truly immersive virtual reality (VR) environment and to allow users to interactively operate simulations in the process opens up promising avenues for exploring complex interactions between users and the traffic scenarios being simulated. The rich possibilities of VR are now just starting to be mature enough to innovate several transportation applications. They can provide the desired output for the traffic environment in a true 3D mode. More importantly, users could "step into" the simulated environment and experience it firsthand.

Driven by the increasing need for illustrating complex scenarios for decision making and public outreach, the main objective of this paper is to develop a framework that leverages the recent developments in cloud computing and virtual reality to advance the visualization of MTS. The proposed study distinguishes itself from traditional 2D and 3D implementations in the way that it provides a truly interactive, multi-user, multi-perspective, and multi-mode (M^3) visualization solution for MTS.

Literature Review

The early generation of MTS models often enables the display of traffic entities in 2D computer graphics. For example, the NETSIM simulation model projects network components, vehicle movement, and operational measures in 2D graphics (*7*). Likewise, TSIS/CORSIM is frequently used to simulate traffic operations of arterials and freeways (*8*). Despite being widely used in the design process, these models cannot create inspiring and convincing visualizations to sufficiently convey scenarios to a layperson and therefore have limited success for public participation and dissemination.

In light of the limitations in 2D graphics environments, some studies have made efforts to enhance the display of MTS models by 3D visualizations. For example, Seneviratne used OpenGL API to perform real-time 3D rendering of the vehicles for developing an intermodal terminal simulation model (ITSIM) (*9*). In fact, many off-the-shelf simulation software such as Vissim, Aimsun, and Paramics have presently incorporated a 3D viewer to visualize simulation (*10*, *11*). In addition, their 3D visualizations can often be further excelled by add-in programs. For instance, Nomden et al. improved the visualization of Vissim by importing additional 3D models developed in 3ds Max for the City of Newcastle (*12*). With a more sophisticated configuration, Sun et al. developed a control program using Vissim API to simulate pedestrian crossing behavior in a virtual environment (*13*). Despite the interaction via the avatar, the program limited participants to conduct experiments in a miniature virtual scene with inevitable visual distortions.

At present, few studies have attempted to leverage external VR toolkits and devices to visualize MTS models in more advanced virtual reality. As an example, Zhou et al. extended CORSIM simulation by implementing its 3D visualization with the EON Studio, a PC-based VR software (*14*). Their approach consisted of two key components: (a) creation of 3D models (e.g., roadways and vehicles); and (b) specification of individual vehicle motion in 3D forms. Despite the reconstruction of the traffic flow in 3D, the study was limited in its ability to create realistic paths of vehicles on curves, smooth motion, and accurate representation of vehicle sizes for collision avoidance. Many important elements such as traffic signals and their control were not modeled in the visualization. Meanwhile, the low rendering speed prevented it from achieving smooth animation. Consequently, only audio video interleaved (AVI) movies were generated in the final visualization. This, in turn, disallowed any interactivity the VR package offered. In Chun et al., a visualization environment was created by exporting simulation results of Vissim to VR4MAX, a 3ds Max plug-in (*15*). Hu proposed the integration of pedestrian simulation in AnyLogic and VR to examine pedestrian flow and facility layout at subway stations (*16*). The proposed procedure used the spatial positions of simulated agents saved in XML format file as the input. Then it applied the micro-simulation plug-in (OpenMicroSim) of the UC-win/Road software to incorporate the XML files and created 3D visualization of the pedestrian movement. Essentially, the plug-in only served as a player to exhibit the imported simulation results. Despite the claimed photo-realistic 3D environment, the visual representation of pedestrian motion was not realistic because all the pedestrians must move along predefined navigation paths. Another issue is the efficacy of the developed procedure. Information about the rendering speed was also not reported. Thus, it was not clear whether the proposed procedure could smoothly visualize the crowds at the stations.

To summarize, a major issue of previous MTS 3D visualizations was the insufficient realism caused by the low fidelity of the 3D environments, object movements (e.g., vehicles, pedestrians), or user interactions. In addition, existing MTS visualization tools only allow a single point of view at any time, no matter how many users are viewing the simulation. This poses a serious problem as many microscopic traffic issues cannot be easily discerned from a single viewpoint. Allowing multiple users to view the same traffic scenario from varied perspectives facilitates traffic problem solving, decision making, and public dissemination. Another drawback of existing MTS visualization is that it is always runs on the same computer as the MTS itself, which severely restricts the number of users who can view the MTS visualization at the same time. Although MTS visualizations are commonly recorded as videos, the lack of interactivity in recorded videos reduces the convincingness of the MTS. In the following section, this paper proposes implementable solutions to address these problems.

Methodology

Empowered by the latest advances in computer hardware, software, and networking, this study proposes a novel framework for interactive visualization of microscopic traffic simulations. The proposed framework offers a truly M^3 visualization architecture for MTS. M1: "Multi-user" means that different model users (e.g., engineers, analysts) that intend to explore the simulation models. M2: "Multi-perspective" represents that these users can view the models in different roles (e.g., pedestrians and drivers). M3: "Multi-mode" denotes that the users can apply different types of devices to view the simulation model. The highly interactive and immersive M^3 MTS visualization provides a powerful communication tool to important stakeholders of traffic simulations and facilitates decision making and public dissemination. The following sections detail the system design and key components of the proposed framework.

System Architecture

Different from existing MTS visualizations that run on the same computer as the MTS itself, the proposed M^3 visualization framework employs a distributed architecture based on the client-server model in which the MTS is the server and various visualizations are the clients. The key components of the proposed framework are illustrated in Figure 1. The MTS server hosts the back-end microscopic traffic simulation engine, which can be any commercial, open source, or custom-made MTS platform. The clients are various computing devices that display the 3D visualizations of MTS, which can be traditional personal computers, smartphones, virtual reality devices, or augmented reality (AR) devices. Both the MTS server and clients are connected to the Internet and communicate over it. In the most common configuration, the

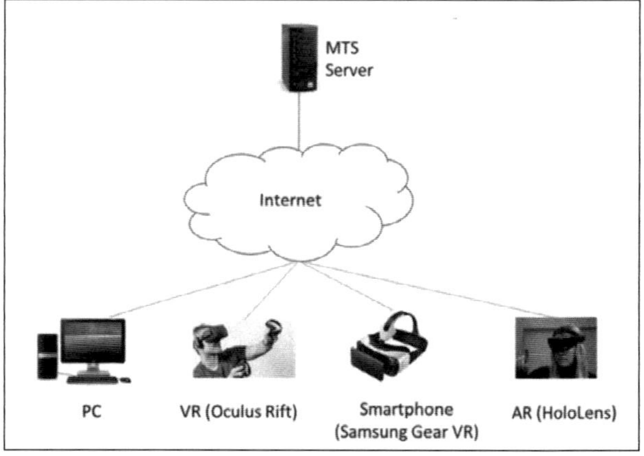

Figure 1. M^3 Visualization system architecture.

server and clients reside on different computing devices. Although they can run on the same computer, the server and the clients still exchange information via network communications. Multiple clients can be connected to the server simultaneously, presenting the same or different views of the same MTS. The server executes simulation as a service (SaaS) and at each simulation time step, sends the vehicle information (ID, position, speed, and bearing) and other traffic information such as traffic signals (IDs, location, and signal states) to the clients.

The clients present a realistic rendering of the 3D environment that contains the road network used by the MTS server and a graphical user interface (GUI) for users to interact with the visualization. The GUI allows users to freely change the viewpoint of the traffic simulation, for example, bird's-eye, pedestrian, driver, and car follower view. Different views facilitate a comprehensive understanding of the traffic simulation from multiple perspectives. The bird's-eye view provides a holistic observation of the overall traffic flow. The pedestrian view is especially important for pedestrian safety studies at various locations such as intersections. The driver view perceives the traffic simulation as a driver inside a moving vehicle and offers better appreciation of the traffic condition that is otherwise impossible or difficult to comprehend, such as weather conditions (e.g., raining), road conditions (e.g., poorly maintained roadways), and lighting conditions (e.g., sun glare). The car follower view allows users to observe the behavior of a vehicle from a third-person view behind the vehicle. Depending on the view selected, users can have further interaction options. For example, users can move as a pedestrian when the pedestrian view is selected and the view will be automatically updated based on users' position and orientation. The proposed framework supports multiple display and input devices ranging from traditional desktop displays to smartphones, VR devices, and AR devices, as shown in Figure 1. The names inside the parentheses are examples of computing device categories, for example, Oculus Rift is one example of a VR device. A user can choose a display or input device that is most convenient or effective for her/him. M^3 visualization of a MTS model provides developers and users an unprecedented capability to visualize traffic simulations and to promote or disseminate simulation findings.

MTS Server

MTS models will be deployed as the server in the proposed framework. The server is a core component of the framework as it provides very high fidelity in both the representation of traffic networks, and dynamic interactions between different road-users and the traffic environment. Even complex traffic conditions can be modeled in an unprecedented level of detail providing realistic traffic models.

From a visualization perspective, most of the existing MTS models already have built-in 2D and/or 3D display functions, for example, Vissim, Paramics, and Aimsun. Despite the differences in visualization procedures, fundamentally these MTS models need to import external 3D objects into the simulation network. For a non-professional user, the 3D representation helps them understand the complex scenario. However, compared with professional 3D graphics tools, the manipulation of the imported 3D objects and their coordination with the entire simulation environment are very restricted. The integration of the 3D objects with the road network is still rather primitive, for example, only basic rendering is utilized without realistic visual effects, such as advanced lighting, specular reflection, and shadows. In addition, existing MTS models restrict users to see a single view at a given time period on the same computer. Although few MTS models (e.g., Vissim) support driver view and pedestrian view, their sights are always fixed in the forward direction. Users cannot freely observe surroundings like a real driver or pedestrian.

To make the proposed framework more adaptable, the selection of MTS models should not be limited to a specific simulation package. Promisingly, our proposed visualization framework does not need to inherit any built-in visualization functions from existing MTS models. Thus, this provides users more flexibility to select a server from a variety of off-the-shelf simulators (e.g., Vissim, Paramics, Aimsun, SUMO, CORSIM, etc.). The minimum requirements for an MTS model to be able to function as a server in the proposed framework include: (a) the capability of generating vehicle trajectory profiles; and (b) an interface to access the traffic control information. The simulation model is not required to have very high resolutions (small time steps, i.e., 0.1 s) as additional interpolation techniques can be applied to smooth vehicle movement if a small time step is not allowed in the model.

Interactive M^3 Visualization for MTS

MTS models simulate traffic scenarios at the individual vehicle level. Realistic rendering of the environment and entities involved in the MTS are critical to effectively convey the findings to a wide range of stakeholders. Existing MTS visualizations do not fully and accurately depict the details of traffic scenarios, for example, no existing MTS visualization provides a realistic first-person view of the traffic situation as a driver. Some traffic situations can only be appreciated from a first-person perspective. For example, the strong sun glare when driving eastward in the morning is a safety hazard that can only be perceived from a first-person view, but not from a top view or bird's-eye view. Other driving conditions such as sight distance and blind spot have similar complications. Visualization of the MTS should take advantage of the latest developments in computer graphics hardware and software, such as graphics processing units (GPUs) and various shaders, and generate a virtual environment of the road network to be as realistic as possible, for example, with shadows and reflection, to present the details of MTS and communicate findings.

Although MTS visualizations can be implemented using traditional display and user interaction technologies, for example, LED monitors, and keyboard/mouse inputs, the newest development of VR and AR in the last few years provides more immersive and natural user experiences that are especially useful for MTS visualization. Several market-leading VR and AR devices are shown in Figure 2. Oculus Rift, shown in Figure 2*a* and developed by Oculus, was the first consumer VR device that generated broad market impact. Facebook acquired Oculus for $2 billion in 2014, starting the trend for the latest investments in VR. Oculus Rift incorporates rotational and positional tracking. However, the positional tracking offered by Oculus is limited with regards to space. HTC Vive, shown in Figure 2*b* and manufactured by HTC, tracks user's position and orientation with a tracking space of 4.6 m × 4.6 m. Base stations (named "Lighthouse" and the two boxes in Figure 2*b*) are used to track user movement with sub-millimeter precision. Both the Oculus Rift and HTC Vive are driven by personal computers that meet specific minimum hardware requirements. The Gear VR, shown in Figure 2*c*, was developed by Samsung. It is different from the Oculus Rift and HTC Vive in that a Samsung smartphone is used as the headset's display and processor, while the Gear VR controls the field of view and uses an inertial measurement unit (IMU) for rotational tracking. This device does not include positional tracking but is more accessible because of its low price compared with the rest of the devices. Microsoft HoloLens, shown in Figure 2*d*, is an AR device that produces an environment in which the real and virtual world are merged. This device allows the user to project 3D models in the real world to visualize and edit them. User interactions are performed through the users' gestures and voice. Although this device does not offer an immersive view of the scene, it produces a hologram that can be positioned and scaled in the real world.

VR and AR devices offer a more immersive and natural user experience that facilitates better understanding of MTS.

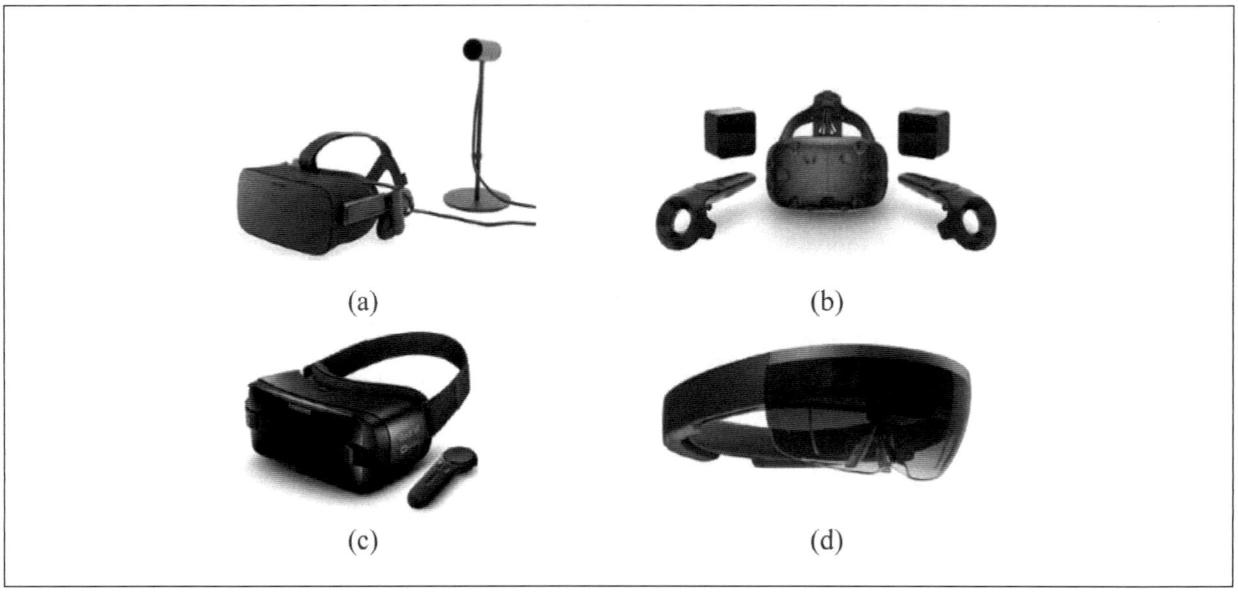

Figure 2. Various VR and AR devices: (*a*) Oculus Rift, (*b*) HTC Vive, (*c*) Samsung Gear VR, and (*d*) Microsoft HoloLens.

For example, devices with positional tracking such as HTC Vive and Microsoft HoloLens can track user movements without any manual inputs and therefore, are especially efficient to be used for automatically updating the pedestrian view following user movements. VR and AR devices, together with the traditional PC displays, offer users multi-mode visualization of MTS. A user can select a mode that is most convenient for her/him and observe the MTS from different perspectives. Allowing multiple users at different physical locations to view the same MTS simultaneously via Internet connections greatly broadens the impact of MTS and facilitates discussion, collaboration, and decision making.

Prototype Development and Demo

A prototype has been developed following the proposed M^3 framework. It was built on a hypothetical city model covering an area of 2.0 miles × 2.0 miles, with approximately 1,000 buildings, road network, and other facilities (e.g., bus stops, trees, signs, etc.) that are typically present in an actual city. It should be emphasized that users can freely replace it by any city models of interest. The following sections provide the detailed description of the prototype development and demonstration. For the prototype, the MTS server currently runs on a laptop with the following configuration: (a) Intel® Core™ i7-3540M CPU @ 3.00GHz; (b) 8.00 GB RAM; and (c) Windows 7 Professional 64-bit Operating System. The modeled simulation network serves approximately 2,400 vehicles per hour. The simulation time step was set as 1/6 second. Users can adjust the demand and time step according to their own needs.

Selection of MTS Server

In this prototype, Paramics (Parallel Microscopic Traffic Simulator) has been used as the MTS server. Paramics is a suite of high-performance MTS tools originating from the Edinburgh Parallel Computing Center in Scotland (*17*). Specifically, its modeler is used in this study. It has a built-in car-following model and lane-changing model that imitate the psycho-physical feedback and gap acceptance behavior of various vehicles stochastically. Apart from the fundamental modeling and analysis features, the most important attribute for us to consider Paramics in this study is its customizable application programming interface (API). As the default traffic models and functions may not always satisfy needs, its API allows modelers to access and develop customized components. More importantly, it allows the simulation model to interface with many external programs and toolkits easily. This is critical for the present study because the simulation model, as a server, needs to communicate and exchange data with various visualization platforms in real time.

Virtual Environment Development

To create a virtual environment (VE) at the clients' end, Unity (more details can be found at https://www.unity3d.com), one of the leading game engines, has been used. Unity is a fully integrated, cross-platform professional game engine, equipped with a wide array of capabilities for game development, such as advanced lighting, animation, game physics, and artificial intelligence. A set of advanced rendering techniques are utilized in the VE, for example, shadow

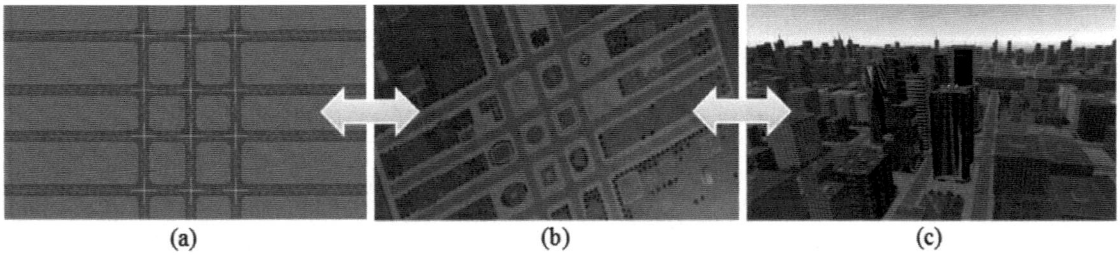

Figure 3. VE development: (*a*) network in Paramics, (*b*) 2D city model, and (*c*) 3D city model.

maps and environment maps. The target road network in the city model matches the node-link road network in the Paramics model. It is imperative that the geometry of the road network in both models aligns appropriately.

Figure 3 demonstrates the development of the virtual environment. The process is reversible, depending on the initial model at disposal.

Other than the static objects in the VE, creating accurate movement of the vehicles is quite indispensable. Unlike any post-simulation visualization, the movement of vehicles in the VE is synchronized with the real-time trajectory profiles of each vehicle. Paramics shared the profiles with the VE through a customized API. Other than the position coordinates, the vehicle bearing information from the simulation model was also shared to determine the orientation of vehicles while moving. In almost all MTS models the vehicle turning is unrealistic because it snaps from one position to another when changing lanes or making a turn in a given time step. To remove the snapping effect, spherical linear interpolation (Slerp) was performed for vehicle rotation at the clients' end, achieving a smoother turning movement.

As the flow of traffic is often regulated by traffic signals in a network, signal control is also implemented for signalized intersections in the VE. This is achieved by linking the signal timing plans modeled in Paramics to update the traffic lights in the VE. Thus, the traffic flow strictly follows the regulations of the signals in the VE.

Distributed Multi-User Visualization System

To test the feasibility of the distributed client-server framework, Paramics was turned into the MTS server by incorporating networking communications using TCP/IP protocols. Each virtual environment visualization client connects to the server using the server's domain name or IP address. The server does not need to know the clients' IP. Up to five clients in different networks were connected to the server in this prototype. Figure 4 demonstrates one experimental configuration.

Figure 4. Multi-user visualization system to support multiple distributed clients simultaneously.

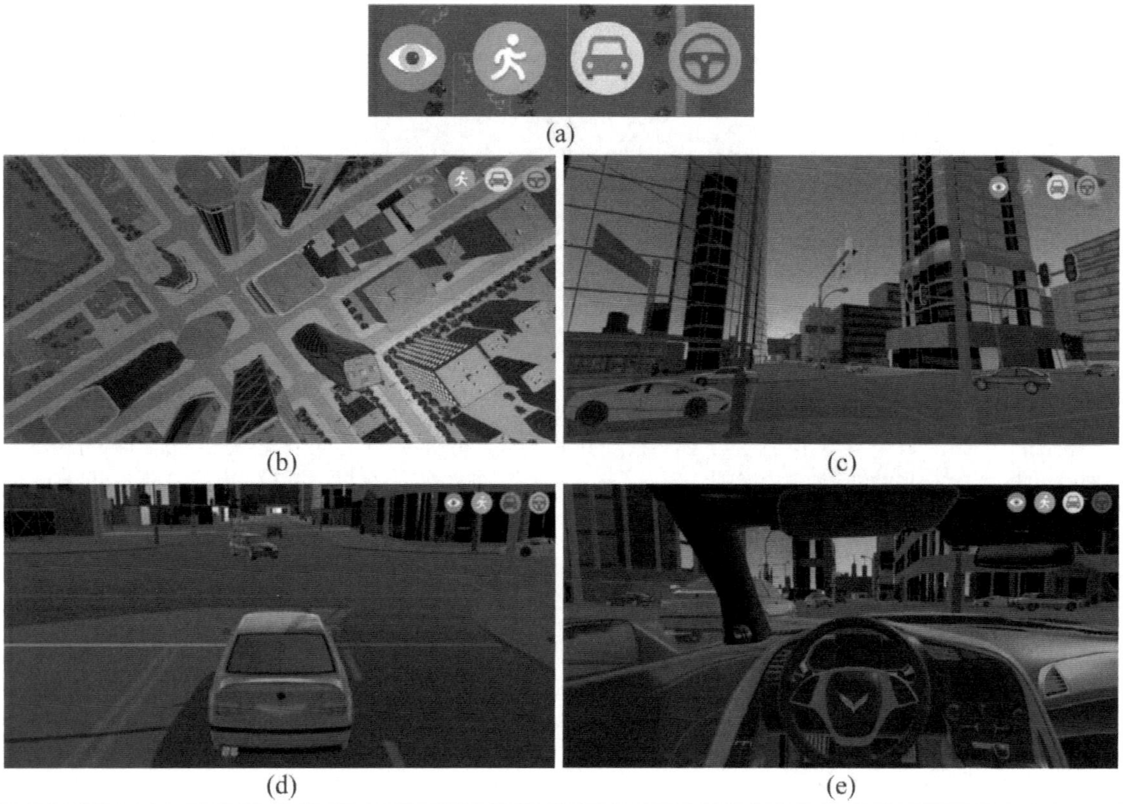

Figure 5. Multiple perspectives of simulation: (*a*) view menu, (*b*) bird's-eye view, (*c*) pedestrian view, (*d*) car follower view, and (*e*) driver view.

Note that the three clients were connected to the MTS server through a different wireless network. The road network in the MTS server shows the simulation model. Each client can freely choose to navigate through a scenario of their interest. For example, Client 1 follows the movement of a turning vehicle. Client 2 examines an intersection while standing nearby the waiting area of a crosswalk. These demonstrations suggest that users in different locations can simultaneously visit the server and "step into" the simulation visualization remotely, which confirms the success of the proposed distributed multi-user visualization framework for supporting project collaboration or public outreach.

Multiple Perspectives

Other than zooming and panning, each user is offered the capability to explore the visualized environment from different perspectives. For the current prototype, four different perspective modes are supported, including bird's-eye view, pedestrian view, car follower view, and driver view. A user can easily select a view using the menu as shown in Figure 5*a*. The buttons from left to right in the figure represent bird's-eye view, pedestrian view, car follower view, and driver view. In the desktop version of the visualization, the views can be selected by clicking one of the four buttons placed in the top-right corner of the screen. In the VR/AR version of the visualization, the users have the ability to pop-up a user interface with the same four buttons for selection.

By default, the visualization starts in bird's-eye view (Figure 5*b*) in which the user can see the overall network from an overhead position. The view can be in an orthographic mode or in a perspective mode. In this view, the user can zoom and pan. When the user selects pedestrian view, a pedestrian is generated and the viewer can see through the pedestrian's eyes (Figure 5*c*), walk around, select vehicles, look up-down-around, and so forth. To be able to follow a car, the user selects car follower view and can then click on any one of the vehicles (Figure 5*d*). The user's view will start to move along with the vehicle from a distance and can look around freely. The same is true for the driver's view (Figure 5*e*).

Multi-Mode Visualization

One of the goals of this project is to offer a multi-mode visualization so that users can visualize the MTS from using devices. To this end, we adapted the visualization to the VR

Figure 6. Demonstration of the multi-mode interactive visualization (*a*) desktop; (*b*) HTC Vive; (*c*) Oculus Rift; (*d*) Gear VR; and (e) HoloLens.

and AR environments, which offer immersive views of the scene. The implementation of the scene in virtual reality was deployed using a HTC Vive, an Oculus Rift and a Gear VR with a Samsung Galaxy Note 4, while the Microsoft HoloLens headset was used to implement the scene in augmented reality. Figure 6 shows the final visualization experiments. Other than the conventional desktop view (see Figure 6*a*), the proposed framework provides users an immersive environment using any of the VR and AR devices (see Figure 6, *b*–*e* for examples).

A crucial aspect in the adaptation of the scene to the VR and AR environments was the change of users' interface and interactions with the scene. In the desktop applications with traditional displays, user interfaces are commonly implemented in the screen space, and a user interface (UI) is usually displayed at a fixed location on a screen, for example as shown in Figure 5, and users interact with the UI with a mouse or keyboard. However, screen spaces no longer exist in VR/AR applications as there are no input devices to interact with such spaces. Instead user interfaces are implemented directly in the world space in the VR/AR application. In this prototype, user interfaces are implemented in a panel that has a fixed distance from the users' eye and can be enabled or disabled from a button on the controller (see the left side of Figure 6*d* for an example). The user has the ability to select a view using the controllers in the HTC Vive, the remote in the Oculus Rift, or the touchpad in the Gear VR. In the case of the Oculus Rift and the Gear VR, the user has to gaze with his/her head to the proper button to select it, as the devices do not include a positional-tracking controller. Figure 6*d* illustrates this selection process, in which the user attempts to select the pedestrian view. Another interaction with the scene is the movement of the users in the first-person view. In this case, we implemented a teleport system with the controllers in the HTC Vive so that the users can move to a point that they select (Figure 6*b*). Users will then see a similar scene to the one on the TV monitor but in a 3D mode through the HTC Vive. Similarly, in the Oculus Rift, the user has the ability to move using the arrows on the remote controller (Figure 6*c*). In the case of the adaptation of the scene to the Microsoft HoloLens, customized hand gestures have been designed for operations such as zooming, panning, selection, rotation, and so forth. The scene can be conveniently demonstrated in any environment using HoloLens, for example, a large auditorium in Figure 6*e*. More importantly, the user can move as a real pedestrian in a 1:1 AR environment without any manual input. None of current microscopic simulation models have such visualization capability.

Conclusion

Visualization is frequently used by traffic professionals for analyzing microscopic traffic simulation models. In this contribution, we presented a framework for enhanced visualization of MTS in a truly 3D virtual environment. The proposed M^3 visualization framework was established by developing a distributed application structure that networks traditional MTS models with different virtual/augmented reality platforms. The method was based on the replication and enhancement of the key simulation components, including the road network, vehicle trajectories, traffic control, and so forth, to create a more realistic 3D environment in a 1:1 scale. A prototype that implemented the M^3 visualization framework was developed and demonstrated. It used Paramics as the server to test the multi-user, multi-perspective, and multi-mode visualization of a typical city traffic model.

The demonstration of the M^3 visualization framework through the prototype provides several valuable implications concerning the visualization of MTS models. On the one hand, it suggests that the advanced visualization of conventional traffic simulation models such as Paramics and Vissim can be favorably accomplished by leveraging the latest VR/AR technologies. Rather than relying on the built-in 3D environment, the MTS visualization in actual scale can be empowered by an immersive virtual environment implemented in various VR/AR platforms. On the other hand, it shows that the visualization of MTS models can be realized in a distributed network. Multiple users can remotely explore a simulation model running on a modeler's computer simultaneously. This facilitates disseminating of simulation projects to the public in real time without using static images and/or recorded videos. Furthermore, it suggests that different users can interactively explore a simulation model from different perspectives concurrently. Unlike the fixed view in traditional MTS visualization, users equipped with traditional displays and/or different VR/AR devices can be granted more degrees of freedom and roles to navigate through a simulation model. This offers stakeholders an unprecedented ability to visually examine complex traffic scenarios of their interest. For example, it can be used to demonstrate the complicated design of the continuous flow intersections that have been frequently implemented in the U.S.. With the M^3 visualization, both drivers and designers will have the chance to explore the concept design in a more close-to-reality context. This will help leverage the existing simulation models developed by designers/consultants to improve the design of the new infrastructure. Other than some 3D models and simulation models, the need for additional data collection effort is limited.

Despite new efforts to visualize microscopic traffic simulation models, more research should be performed to further help excel the M^3 visualization framework. Clearly, there is need for a centralized web interface so that everything (or almost everything) can be conveniently run in a web browser. The current framework requires each user to have a client application installed before connecting to the server. The web interface will allow them to establish the connection and execute the visualization more easily. Another aspect that deserves more effort is to extend the framework for supporting other commonly used MTS models such as Vissim and Aimsun. As each MTS model has different features (e.g., data storage, API, etc.), the framework needs additional customized modules for accommodating each type of microscopic simulation models. Furthermore, continuing development of the framework will allow multiple users to communicate and collaborate with each other in real time. This can be accomplished by adding a hierarchical structure of users (e.g., administrators, presenters, ordinary users, etc.) to perform different tasks. In addition, (semi-)automated processes to develop the virtual environment are desirable, especially for large networks. For example, registering the road network in the 3D virtual environment is time consuming. Efficient ways to generate the road mesh will be very helpful. Last but not the least, the proposed framework can be readily expanded to supporting human-in-the-loop simulations, in example, some users can control the vehicles as drivers in the MTS simulation. As there will be potential for motion sickness using the VR/AR headsets, it will be necessary to study this issue in future research. Our current work did not involve very complicated motions. When tasks such as vehicle controls are involved, it will be important to examine the tolerance of motion sickness by drivers. Several of these future efforts are already underway in our team.

Acknowledgments

We appreciate the support of the MSVE at ODU, especially Dr. Rick McKenzie in MSVE for the great help. The contents of this paper only reflect views of the authors and do not necessarily reflect the official views or policies of any sponsoring agency.

Author Contributions

The authors confirm contribution to the paper as follows: study conception and design: Yuzhong Shen, and Hong Yang; data collection: Hong Yang, Yuzhong Shen, Mahmud Hasan, Daniel Perez Ibanez, and John Shull; analysis and interpretation of results: Hong Yang, Yuzhong Shen, Mahmud Hasan, and Daniel Perez Ibanez; draft manuscript preparation: Hong Yang, Yuzhong Shen, Mahmud Hasan, and Daniel Perez Ibanez. All authors reviewed the results and approved the final version of the manuscript.

References

1. Chien, S. I. J., D. G. Goulias, S. Yahalom, and S. M. Chowdhury. Simulation-Based Estimates of Delays at Freeway Work Zones. *Journal of Advanced Transportation*, Vol. 36, No. 2, 2002, pp. 131–156.
2. Lee, J., and B. Park. Development and Evaluation of a Cooperative Vehicle Intersection Control Algorithm under the Connected Vehicles Environment. *IEEE Transactions on Intelligent Transportation Systems*, Vol. 13, No. 1, 2012, pp. 81–90.
3. Le Vine, S., A. Zolfaghari, and J. Polak. Autonomous Cars: The Tension between Occupant Experience and Intersection Capacity. *Transportation Research Part C: Emerging Technologies*, Vol. 52, 2015, pp. 1–14.
4. Bloomberg, L., and J. Dale. Comparison of VISSIM and CORSIM Traffic Simulation Models on a Congested Network. *Transportation Research Record: Journal of the Transportation Research Board*, 2000. 1727: 52–60.
5. Sbayti, H., and D. Roden. Best Practices in the Use of Micro Simulation Models. Presented at American Association of State Highway and Transportation Officials (AASHTO) Standing Committee on Planning, AECOM, Arlington, Va., 2010.
6. Hidas, P. Modelling Lane Changing and Merging in Microscopic Traffic Simulation. *Transportation Research Part C: Emerging Technologies*, Vol. 10, No. 5, 2002, pp. 351–371.
7. Lieberman, E. B., and B. Andrews. The Role of Interactive Graphics When Applying Traffic Simulation Models. *Proc., Winter Simulation Conference Proceedings*, New Orleans, La., IEEE, New York, 1990, pp. 753–758.
8. Garrick, N. W., P. Miniutti, M. Westa, J. Luo, and M. Bishop. *Effective Visualization Techniques for the Public Presentation of Transportation Projects*. Report No: NETC 48. University of Connecticut, 2005.
9. Seneviratne, P. N. *Simulation and Animation Model for Planning and Designing Transit Terminals*. IDEA Programs, Transportation Research Board, Washington, D.C., 2005.
10. Boxill, S. A. *An Evaluation of 3-D Traffic Simulation Modeling Capabilities*. Report No: SWUTC/07/167621-1. Texas Southern University, Houston, 2007.
11. Papageorgiou, G., P. Damianou, A. Pitsillides, T. Aphamis, D. Charalambous, and P. Ioannou. Modelling and Simulation of Transportation Systems: a Scenario Planning Approach. *Automatika: Journal for Control, Measurement, Electronics, Computing & Communications*, Vol. 50, No. 1–2, 2009, pp. 39–50.
12. Nomden, H., D. Greenwood, M. Horne, T. Muizelaar, and E. M. Thompson. Traffic Simulation in 3D World. *Northumbria Working Paper Series: Interdisciplinary Studies in the Built and Virtual Environment*, Vol. 2, No. 1, 2009, pp. 26–38.
13. Sun, J., Y. Yang, and H. Wang. Development and Application of a Simulation-Enhanced Platform for Pedestrian Crossing Behaviors Experiment. Presented at 90th Annual Meeting of the Transportation Research Board, Washington, D.C., 2011.
14. Zhou, B., J. Blum, and A. Eskandarian. Virtual Reality Visualization of Microscopic Traffic Simulations. *Transportation Research Record: Journal of the Transportation Research Board*, 2005. 1937: 159–166.
15. Chun, W., C. Ge, Y. Liu, and M. Horne. Virtual-Reality Based Integrated Traffic Simulation for Urban Planning. *Proc., 2008 International Conference on Computer Science and Software Engineering*, Hubei, China, IEEE, New York, 2008, pp. 1137–1140.
16. Hu, M. A High-Fidelity Three-Dimensional Simulation Method for Evaluating Passenger Flow Organization and Facility Layout at Metro Stations. *Simulation*, Vol. 93, No. 10, 2017, pp. 841–851.
17. Cameron, G. D., and G. I. Duncan. PARAMICS—Parallel Microscopic Simulation of Road Traffic. *The Journal of Supercomputing*, Vol. 10, No. 1, 1996, pp. 25–53.

The Standing Committee on Visualization in Transportation (ABJ95) peer-reviewed this paper (18-03025).

Article

TRR
JOURNAL OF THE TRANSPORTATION RESEARCH BOARD

Transportation Research Record
2018, Vol. 2672(44) 72–81

DOI: 10.1177/0361198118792334
journals.sagepub.com/home/trr
SAGE

Simulation Games to Study Transportation Issues and Solutions: Studies on Synchromodality

Ioanna Kourounioti[1], Shalini Kurapati[1], Heide Lukosch[1], Lóránt Tavasszy[1], and Alexander Verbraeck[1]

Abstract

In this paper we discuss the application of simulation gaming to study the behavior and decision making of stakeholders when confronted with complex transportation problems. The problem we tackle is synchromodal transportation. Synchromodality requires the vertical and horizontal collaboration of stakeholders in all the levels of decision making. To facilitate this, we develop four games designed in a way that meets the needs of decision makers in each level. We present both board and digital games and the results from the first gaming sessions with Dutch supply chain and logistics professionals.

Synchromodality is the "vision of a network of well organized and interconnected transport modes, which together cater for the aggregate transport demand and can dynamically adapt to the individual and instantaneous needs of transport users" (*1*). The required level of synchronization demands horizontal and vertical integration in the design and operation of intermodal transport and the provision of the necessary infrastructure to ensure the on-time and reliable delivery of goods (*2*). The complex nature of the freight transport system can be attributed to the existence of numerous interactions between numerous economic and political actors (*3*), which hinders the orchestration of transport modes and services. Conflicting and complicated decisions are made in all the freight transportation layers for enabling efficient services, infrastructure and assets (Figure 1, adjusted from Schoemaker et al.(*4*)).

One of the key issues that hinders the coordination between the different actors of the supply chain is the lack of information exchange between the actors both in the horizontal as well as in the vertical level. This lack of information exchange is partly due to the lack of collaboration between the stakeholders (*5*). To achieve the optimal level of synchronized services it is necessary to achieve the optimal collaboration between all the stakeholders in the supply chain (*6*). On the other hand, logistics and freight transport in general is a very complex and competitive environment (*3*). In order to bring the stakeholders together in such a challenging environment to understand the various issues and search for solutions we use simulation gaming as our main research method. The key merit of simulation games is that we can study the behavior and decision making of stakeholders, who would otherwise not interact in a real-life setting, in a safe, controlled and relaxed setting. In our studies we mainly used simulation games to study the issues faced by different stakeholders to arrive at synchromodal solutions for efficient transportation. We also apply games to explore solutions and policy measures to counter the discovered issues. In this paper we describe the five games developed for the stakeholders in each decision-making layer. The games presented below are developed in the context of a research project that uses them as a tool to study the behavior of stakeholders towards the complex issue of synchromodality.

This paper is structured as follows. First, we give a brief overview of the key literature findings in the area of the application of simulation gaming as a research method in complex systems such as transportation networks. Then we provide a detailed presentation on the games developed to address the synchromodal transportation issues. We discuss the first results from the gaming sessions and conclude the paper by presenting the points of future research.

[1]Technology, Policy and Management, Delft University of Technology, Delft, Nederland

Corresponding Author:
Address correspondence to Ioanna Kourounioti: ikouroun.sfc@gmail.com

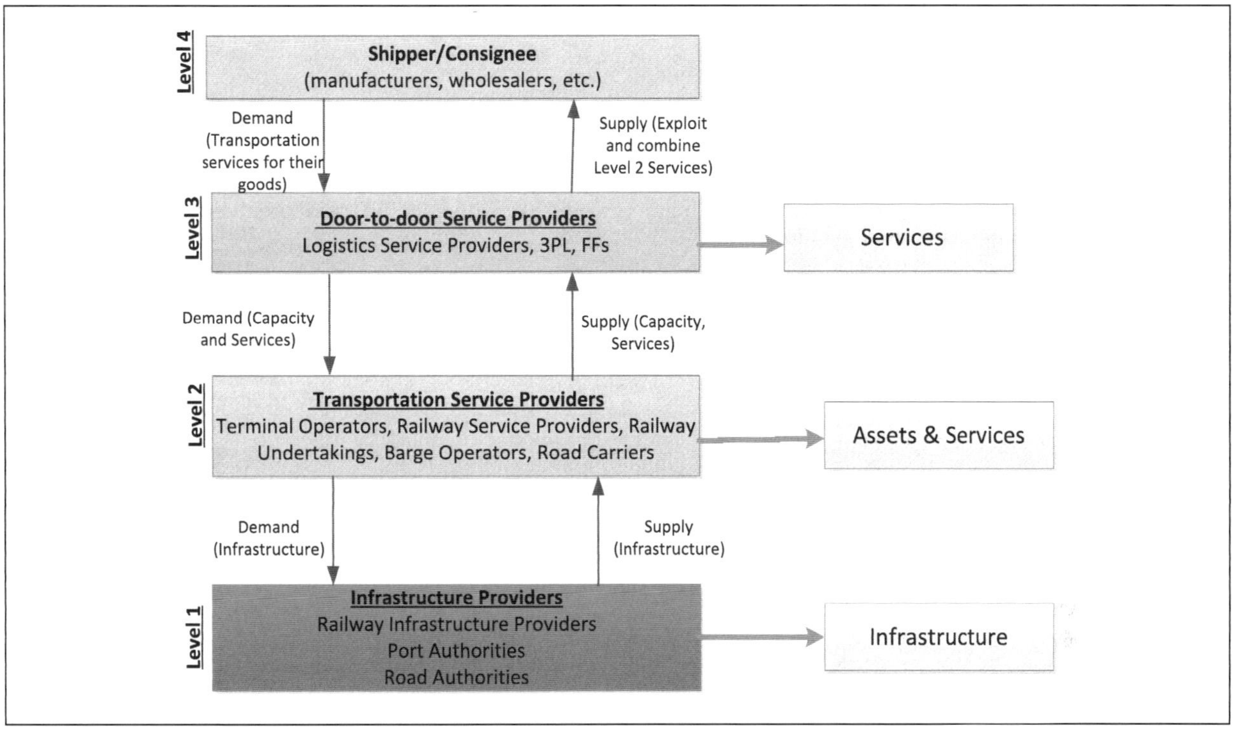

Figure 1. The TRAIL layer model for freight transport.

Background on Simulation Gaming as a Research Method

The modern era of simulation games began in the 1950s through the integration of military/war games, computer science, and operational research (*7*). Simulation games represent a real system at a certain level of abstraction, where participants have to abide by certain rules to win the game by managing their limited resources of money, time, and influence within the game (*8*). Making an analogy to the characteristics of a standard research method, the research problem is the game objective, the research environment is the game itself, and the experimental setup is the game session. Data collection and data analysis can be done from observation and game logs. In the words of Greenblat (*8*), "The design of a simulation game itself is a systematic translation of understandings into an operating model, and the subsequent examination of the model through observation of play can lead to a refining of theoretical formulations and consequently to a higher level of social scientific understanding." Klabbers (*9*) claims that simulation gaming is an effective research method to study complex systems.

This vision was earlier shared by Duke (*10*), who proposed simulation gaming to study issues that are complex, future-oriented, and of a systems nature. The widely used research methods in the field of supply chain, transportation, and logistics are case studies, questionnaires, action research, modeling and simulation, and as a relatively new method, simulation gaming (*11–13*).

Simulation gaming addresses questions of human behavior in the context of organizational action, social change and technical development (*16*). It is well equipped to perform multidisciplinary research that integrates different perspectives, concepts, theories, data, information, methods, techniques, and tools from two or more disciplines to understand and solve problems for which solutions are beyond the scope of a single discipline (*16*). This approach is particularly suited to transport applications in which stakeholders with different perspectives and different information management regimes and solutions are present. Simulation games have been used for transport applications in works of several authors (*11, 17–19*). The works of these authors illustrate cases of how to design and apply simulation gaming for transportation issues. While Meijer (*11*) proposes a solution on a strategic level, Kurapati et al. (*17*) suggests the use of short games on operational level in transportation. Kurapati et al. (*18*) discusses the use of games for the governance of the transport system as a whole, and awareness for synchromodal transport solutions in particular. The cases show that simulation gaming can address challenges and issues of transportation on all three levels; strategical, tactical, and operational.

Table 1. Prominent Research Methods in the Field of Supply Chain, Logistics and Transportation

	Case studies	Surveys	Action research	Modeling and computer simulation	Simulation gaming
Pros	In-depth study into real world situations. Direct observations of real actions and communication of actors.	The ability to gather a large number of responses while giving little or no disturbance to the actual behavior of the respondents.	The involvement of actors in all stages of research bridges the gap between research and application.	Unlimited number of experiments can be conducted with different scenarios.	Observation of actual actions and behavior of participants is possible. The experiments can be repeated in a safe and controlled environment.
Cons	Low repeatability and low ability to generalize due to changing contexts and contextual bindings.	Truthfulness of responses, and involvement of participants cannot be known. Questions may be interpreted differently by different respondents.	Lack of standardization in research procedures and processes. Time-consuming. Loss of control of the research environment due to varying commitments of the actors.	Real data and observations are unavailable. Tacit knowledge of humans is hard to model.	A part of reality of the actual environment has to be compromised to enable playability of the related game. Complex, time consuming, and expensive.

Note: Adapted from Popper (*14*), and Summer and Summer (*15*).

We will introduce the four games developed in our research study on synchromodality in the following sections.

Simulation Games on Synchromodality

Synchromodality is "the optimal, flexible and sustainable use of different transport modes in a network under the control of a logistics service provider, in such a way that the customer (client or freight forwarder) obtains an integrated solution for his (hinterland) transport." (*20*). To enhance the complex vertical and horizontal collaboration and information provision needs inherent with synchromodal transport we developed a game addressing the key decision makers in each level of the layered freight transport model as it can be seen in Figure 2.

Specifically to address the needs, problems and preferences of the stakeholders in the third layer, the SynchroMania game was developed. We designed two board games to address the issue of rail building and cooperation between rail service providers: Rail Cargo Challenge Rotterdam, and Rail Cargo Challenge Amsterdam. For the decision makers on the first level we developed the Modal Manager game in which infrastructure managers use information provision as a way to control traffic and increase the efficiency of the network. Finally, we are in the process of developing the Shipper's Game that investigates the shippers' preferences on the design of the synchromodal transport product. Some of the games have already been presented (*17*, *18*, *21*, *22*) but in this paper we focus on describing the approach of developing games that address the decisions made by the stakeholders in every layer of the logistic chain. The rest of this section provides a more detailed description of the games.

Modal Manager Game

The Modal Manager game is a so-called "microgame" and consists of different missions, representing a network of road, rail and barge transport routes together with the various intermodal terminals in a hypothetical transport corridor in the Netherlands presented in Figure 3.

The learning goal of the game is to investigate how infrastructure providers can apply different information management strategies to inform, advise and control traffic in the network to increase performance and network users' satisfaction. The objective is to raise the awareness of inframanagers; on using information and to understand how to apply different information provision strategies. In the game the player assumes the role of an infrastructure manager (inframanager) and has to deal with four logistic service providers (LSPs), A, B, C and D, who are automated agents. Each of the LSPs is modeled such that they make route choices based on their respective priorities: A is known for reliability, B for sustainability, and C for being cost effective. D introduces an element of randomness without a clear priority. The game can be configured to accommodate several points of origin and destinations for the containers. Container traffic enters the network from one specific origin and has a specific destination (*18*). During the

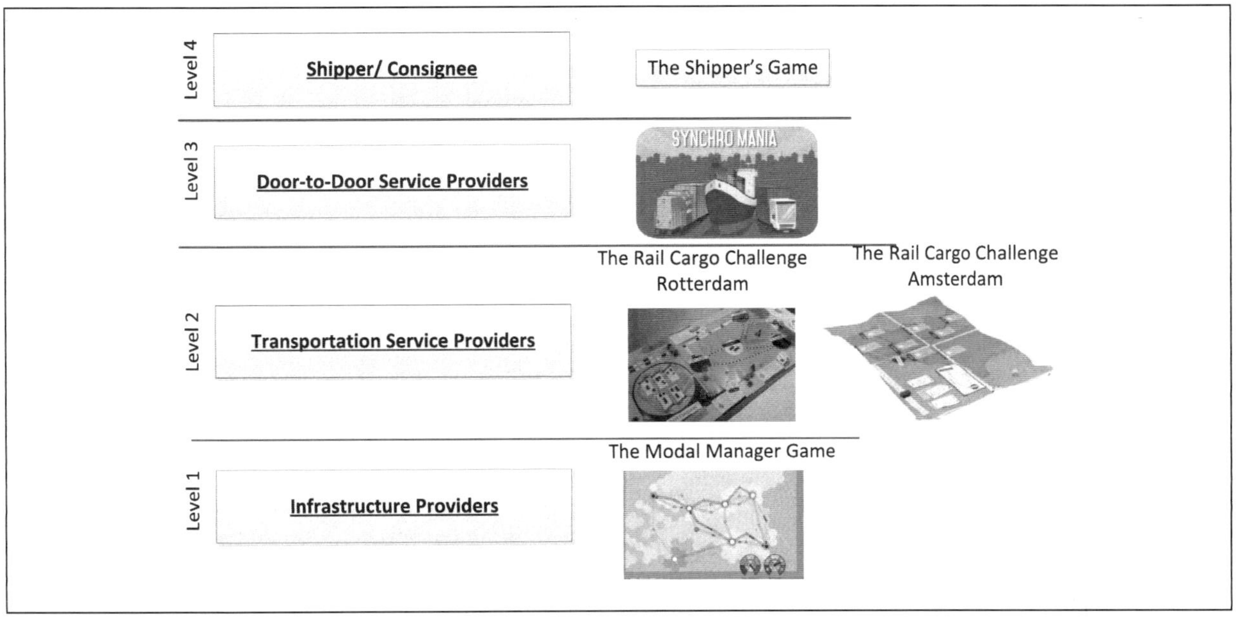

Figure 2. Games per layer.

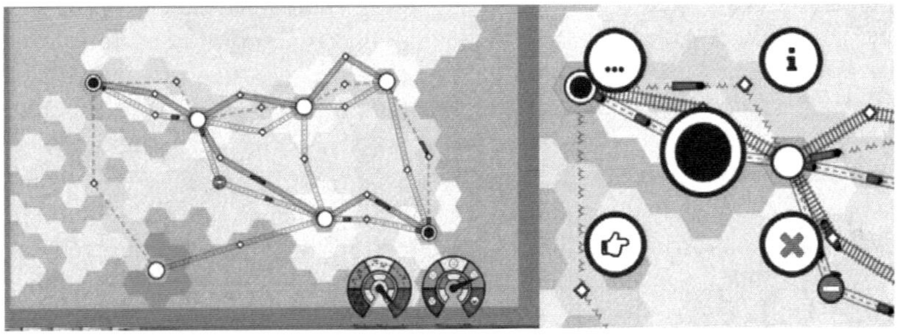

Figure 3. Screenshot and commands of the Modal Manager Game.

game the stress on the capacity of network as well as planned and unplanned disruptions are communicated to the player. Specifically, the player knows the position, the duration and the time that a planned disruption will occur. Unplanned disruptions pop up randomly in the network and their duration is communicated to the player. Projections of expected congestion levels in all the network links, starting from a free-flow and going up to a totally congested link, are also indicated to the players. At the end of the game the players must ensure that all the containers are delivered to the final destination on time. The goal of the game is to pay attention and maintain a good happiness level of the LSPs (the right metric in the right image in Figure 3) and a high overall network performance (the left metric in the right image in Figure 3). This introduces a dilemma to the player to balance network efficiency and happiness levels. The tools at the player's disposal are the different strategies for the provision of information. Specifically, the measures at the players' disposal are (also illustrated in Figure 3, left image):

- **Inform**: The inform option corresponds to a network information system that provides generic information about disruptions displayed and accessible to all LSPs. Based on this information the LSPs can decide whether to take an alternative route or continue on the same route. The interpretation of the information is subjective, and its consequences are unknown to LSPs.
- **Advice**: With this option the inframanager shares only relevant information to relevant LSPs by

explaining the consequences of the final choice of the LSP. Inframanagers must pay attention to the amount of the advice they provide in order not to create information overload and decrease the dissatisfaction of the agent LSPs.

- **Direct**: The inframanager enforces a certain route to the LSP. To ensure the optimal utilization of the direct command the inframanager has to assess the current situation of the network and predict the future congestion levels on the network links.

When providing information, the inframanagers have to make a trade-off between LSP satisfaction and performance of the network. The overall goal of the game is to explore which measures are taken (decision-making of the players), and what affects the game KPIs which are the network performance, measured by the average delays in the network, the number of containers that arrive at their destination on time, and the user satisfaction measured by LSP satisfaction and trust levels. A complex algorithm measures the trust level of LSPs based on the accuracy of information provided by the inframanager. The game has several complexity levels. In the first level of the game, the player controls only one modality within the synchromodal network and operates in a situation which reflects the current situation of managing the network of one mode. In the second level, the player is still in charge of one mode but has to consider the preferences of LSPs while choosing the different measures. In the final game play level, the player is a "super modal manager" who can manage all three modalities.

The game is played by the infrastructure providers of different modalities. The Modal Manager is not designed to be played alone but will be a part of a workshop which starts with a briefing lecture about corridor management by the facilitator. The participants fill in a pre-game questionnaire that collects their demographic information, their expectations from the workshop and their current attitudes towards synchromodal corridor management. Then they play different missions. After each mission, the facilitator invites the participants to reflect on the game play and relate it to real-world applications.

At the end of the workshop the participants are asked to reflect on the following:

1. Reflect on advantages and disadvantages of the current practice of information provision as a way to manage the network.
2. Understand the need to move from operational focus (disruption management) towards a more tactical perspective to balance performance and client satisfaction.
3. Deal with capacity limits through the communication of congestion levels to the network users.
4. Opportunities to communicate and cooperate with LSPs.
5. Opportunities of collaboration with other mode inframanagers.
6. Understand the opportunities and challenges of synchromodality.

Several workshops using the Modal Manager game have been planned together with the infrastructure managers of the Dutch transport networks in winter 2017 and spring 2018.

Rail Cargo Challenge Rotterdam Board Game

The key research objective of the Rail Cargo Challenge Rotterdam (RCCR) is to assess the attitudes and behavior of stakeholders in the freight transport domain with respect to the efficient bundling of containers to be transported to their final destination using rail. The RCCR is a multiplayer table-top board game, set in the environment of a sea port corridor connecting Rotterdam to several destinations in Europe. To construct the new Maasvlakte 2 port extension the port authority had to guarantee a 18% increase in the rail transport share (*23*). The biggest hindrance to this is the dispersed nature of the terminals in the port area. The game was designed in collaboration with the stakeholders from the Port of Rotterdam and addresses the problem of bundling freight from the three terminals in Maasvlakte (*24*). The roles in the game are two competing rail operators who can transport freight using rail by charging a price to shippers. The shippers organize the transport of the container from various container terminals in the port. In the first round of the game, there are three terminals: A, B and C.

Each shipper has order cards that denote the number of containers that need to be transported, the terminal in which they are stored, their destination and the time limit for transporting them. The rail operators have to pick up freight from different terminals in the port at a predefined or negotiated price. However, rail operators have limitations in picking up containers and cannot serve all the terminals. Each rail operator throws a dice to determine the number of terminals that the player's train will be able to visit. All rail operators and shippers start with 50 tokens representing money. The rail operators can arrange trains with capacity of 10 containers each. Each train costs 10 tokens. The shippers can make arrangements with the rail operators to pick up their containers from a terminal—the terminal where the container is allocated is indicted on the order card—for a specific price. For each shipment (order card) that is successfully delivered via rail the shipper receives 4 tokens. On the other hand, if a shipment is not sent by rail then it has to be sent by truck to the final destination, but this option is more expensive. The rail operators will benefit most if they can manage to fill up all their trains to full capacity and make sure they can pick up shipments from the terminals as agreed with the shippers. If the dice is in their favor, and they are able to transport all containers as planned on time they receive 4 tokens per shipment. If they fail to do so, then they have to pay additional tokens to truck the shipments. An extra mechanism inserted in the game is the reputation of the port. If shippers and rail operators do not succeed in delivering the containers at their final destination on time then the reputation of the port is reduced and fewer shipment cards are given to the players in the next round. This mechanism was inserted in the game to make players understand the importance of cooperation. If player does not cooperate shipments will not be delivered, the reputation will go down and clients are lost from the port. In the subsequent rounds the situation becomes even more complex when two new terminals open up in the port, that makes the starting position of the containers more scattered across the port, while the rail services and frequency remain the same (*21*).

The game is part of a gaming session at the end of which participants reflect on the opportunities and challenges of rail bundling and on the importance of cooperation to increase the port throughput as well as the demand for the shippers and services of the rail operators. A related game for the Port of Amsterdam will be discussed in the following paragraph.

Rail Cargo Challenge Amsterdam Board Game

The Port of Amsterdam Challenge Game (RCCA) was also designed to address a similar problem for freight bundling in the Port of Amsterdam. The learning goal of RCCA is to raise the awareness of players about the benefits of cooperating to consolidate freight from the different terminals in the port area. The biggest issue regarding rail freight bundling raised by Amsterdam Port Authority is the existence of five different specialized terminals in the port area. Each terminal handles a specific type of cargo and does not collaborate or exchange information with the other terminals. At the moment a shuttle train service operates in the port that reaches all the terminals. Due to the lack of collaboration half-empty trains circulate around the port without a schedule creating traffic issues in rail network. Amsterdam Port Authority expects that coordination between the terminals to bundle freight will increase the efficiency of rail transport within the port and the productivity of the port in general. Therefore, this board game aims at bringing the terminal operators together and raising their awareness about the benefits of collaborating and consolidating transport.

In the game each player acts as a terminal operator who has to store a number of loads in the terminal and ensure that they are transported to their final destination using rail. At the beginning of the game each terminal already has some existing infrastructure and is specialized in a certain type of cargo. Some of the terminals have certificates that permit them to handle either inflammable cargo, chemicals or grains. Each terminal is given a shipper/client card that shows the name of the client, the type of cargo that the client handles, the final destination that the cargo has to be transported to and the profit that the player can make for the delivery of each unit of cargo. In addition, the player is given one of the following roles:

1. **Logistics expert.** This player faces no delays in the train.
2. **Train expert.** This player starts first putting together trains and has a discount.
3. **Marketing expert.** This player is the first one to choose new shippers.
4. **Infrastructure manager.** This player is the first to buy and has discount to purchase new infrastructure.
5. **The lobbyist.** This player is the first to buy and renew certificates and has a discount on the purchase and the renewal of new certificates.

After storing all cargo in the terminals players must start the procedure to transport them to their final destination. They have two options: the first option is the shuttle service which is much more expensive but guarantees the on-time delivery in any destination; or the player can put together a train that can have up to three destinations. Each train has capacity of 10 containers and costs 13 tokens (the train expert pays 10 tokens for the train). The player can collaborate with the rest of the players to fill all the available capacity in the train. The train expert is the first to put together a train while the logistics expert has no delays. After booking trains the players throw the dice for each destination of the train to randomly see if the train is delayed. For each delay an additional token is charged. Then all the transported contracts are reimbursed and players can buy new infrastructure and certificates. Afterwards five new shippers are inserted on the game board. The players choose a new shipper (they keep their old one) and the next round starts. In should be noted here that in the case that a shipper's contract is not delivered then the reputation of the port goes down and one shipper is removed from the board for the next round. The reputation mechanism points out the importance of collaboration between the terminals to serve all the clients and increase the traffic in the port. The higher the number of rounds the more complicated the game becomes since the players have to store all the cargo in

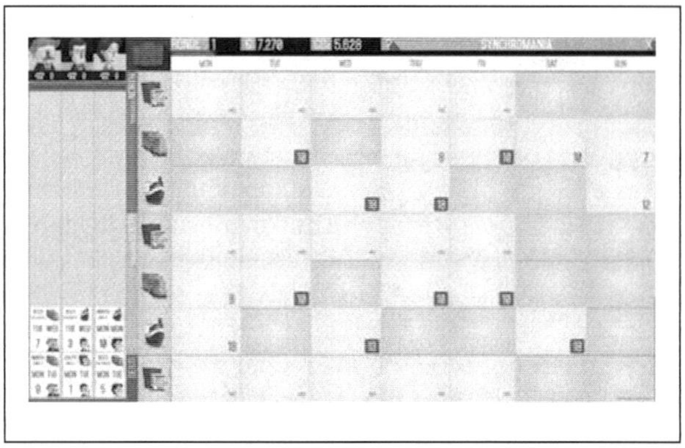

Figure 4. SynchroMania planning screen shot.

their terminals and make sure that they book transport services to deliver their cargo to the final destination on time.

The game is part of a gaming session at the end of which participants reflect on the benefits of cooperating to improve not only the competitiveness of their terminal but also of the Port of Amsterdam as a whole. The fourth game in our study is called SynchroMania and will be discussed next.

SynchroMania

SynchroMania is a digital game that addresses decision makers on the third level (Figure 4) mostly aiming at freight-forwarding companies and LSPs. The objective of the game is to make players understand the complexity of planning orders to the available modes and services to transport all the cargo on time. The long-term goal of the game is to persuade shippers to leave the mode of transport free for the LSP to choose. In the game, players take over the role of the planner and they are asked to optimally place orders, coming from three different customers, into three modes of transport (truck, rail and barge) based on customer's demands, mode's services and space availability. In each round limited time to assign the orders to the specific route and the modality dictated by the client is given. Every round represents a weekly time horizon and a specific time is given to each day. At the beginning of every day new order cards are generated. The mode, the number of containers, the client who places the order, the route (north route or south route), and the pick-up day and the delivery day are written on the card. The player has to plan all the orders and deliver them on time, using the mode and the route indicated on the card. This task is very complicated because there are restrictions on the available capacity of the modes and the frequencies of services. The player has the option to call the client and ask him for flexibility in the characteristics of the shipment. An algorithm generates the possible negative or positive replies of the clients. After the end of every round additional degrees of freedom, in the forms of action cards purchased to increase flexibility, are provided to the players. After a round finishes key performance indicators (KPIs) that calculate costs, CO_2 emissions and customer satisfaction are applied to evaluate players' performance. During the discussion that follows every round the scores are compared, and players discuss their chosen tactics. Based on the insights given in the discussion players can change their tactics in the next round (*22*).

At the end of the game during the debriefing session the players are asked to reflect on (*25*):

- the advantages and disadvantages of synchromodality;
- the difficulties of the planning activities of planners;
- the effect on the work of planners and sales employees;
- the way decisions influence the efficiency in the network.

All the games are part of an on-going research project. We will present our initial findings from the workshops organized using the games so far on their impact on stakeholders in the Dutch transport domain.

Design and Outcomes of Gaming Sessions

The challenges for the development of synchromodal freight transport solutions are due to complexity of the transport system that comprises numerous and often competing stakeholders (*24*). As a first step to engage stakeholders we have developed a series of games that refer to problems faced in the different layers of decision making. All the games are part of a gaming session in which during briefing and debriefing sessions players are enabled to reflect on what they experienced in the game, and to translate this back to alternative solutions. We recorded player attitudes through pre- and post-game surveys, in-game logs, and observer notes.

In the rest of this section we will summarize the key findings from the gaming sessions that were carried out using the simulation games described in the earlier sections.

Modal Manager Game

Two sessions have been carried out using this game with 16 participants. During the briefing of the sessions participants were introduced to the issues of network infrastructure management with a special focus on information exchange and disruption management. In every round the type of information management strategy (informing, advising, directing) influences the likelihood that the LSPs will follow the information. It also affects the ability of the inframanager to maintain the capacity of the network without congestion. Participants were then asked to reflect on the effectiveness of the measures toward efficient traffic management despite the disruptions. They all highlighted the importance of information provision as a means to deal with disruptions (planned and unplanned ones) in the network. They also stressed the innovative nature of information management strategies in the game as a tool to steer the traffic. Participants were very enthusiastic about the game and pointed out that they had fun while they learnt about the importance of collaboration and coordination between the different means of transport to provide a more efficient and less disruptive transportation of goods in the network. The overall refection is that the gaming sessions can successfully contribute to the learning goal of the game.

RCCR Board Game

We organized a game session with 20 participants who were professionals in the Dutch logistics industry at a transport conference held at Tilburg, Netherlands. The session comprised three parts:

1. Briefing and pre-survey—The game master gave a brief introduction on the background of the game and requested the participants to fill in a pre-game survey. In this survey, the participants were asked about their attitudes towards freight bundling together with their demographic information.
2. Game-play—The game play lasted one hour, and two rounds of the game were played.
3. De-briefing and post-survey—During the debriefing participants discussed their post-game experience and filled in a survey to measure the change in attitude of the players related to freight bundling using rail.

The results of the pre-survey state that the 76% of the respondents agreed that rail bundling would increase rail efficiency by decreasing the amount of trains traveling empty or half empty. They also believed it could lead to lower costs and overall improvement of the sustainability of freight transport. The design of an efficient rail network, with optimized and synchronized schedules that permitted the consolidation of freight could increase the benefit of train modal share. According to the respondents, the promotion of collaboration between the different stakeholders had been a critical point. On the other hand, professionals highlighted the challenges they expect to face when consolidating freight. For them, information and data sharing are the biggest hindrances in freight bundling together with legal issues. Bundling is complex because it requires mode coordination, capacity availability, and synchronization of modes and services to align transport containers within the given time constraints. After the end of the session 15 out of 20 participants agreed that the game gave them an insight on rail bundling and helped to reflect on the challenges and opportunities related to freight consolidation. Significant modifications were not observed regarding their opinions on rail bundling They continued to believe that cost reduction, environmental benefits, and higher rail efficiency are amongst the most important advantages of rail bundling. This could be partly attributed to the fact that the participants were already predisposed towards synchromodal solutions since they were managers and executives of transport companies and not involved in the actual operations. On the other hand, the majority of respondents highlighted the importance of designing e-solutions to enhance the provision of essential information for the scheduling of services and to the related legal issues.

RCCA Board Game

The RCCA has many similarities to the RCCR game therefore we designed a similar gaming session for these two games. We organized a session with 15 managers from the terminals in the Port of Amsterdam. During the briefing the game master introduced the game and the issue of cooperation to bundle freight between the terminals in the Port of Amsterdam and participants filled in a pre-game questionnaire on their characteristics and their attitudes towards freight bundling and cooperation between the terminals. The game play consisted of five rounds and in the debriefing players reflected on their experience and filled in a postgame survey. The analysis of the questionnaire data showed that 13 out of 15 participants highlight the importance of communication between the terminals to facilitate freight traffic. From the answers in the postgame questionnaire we observe a switch in the opinion of the players; they all say that they strongly agree about the importance of cooperation. This change of opinion proves the successful design of the game.

SynchroMania Game

SynchroMania is also played within a workshop with briefing and debriefing sessions. During the briefing the game master explains the idea of synchromodality and the game rules. During the different rounds of the game participants compare their scores and discuss their game tactics. During the debriefing the players are given a questionnaire about their characteristics, attitudes and their opinions on the factors that can influence the adoption of synchromodal transport. We have collected data from 59 participants in two different sessions. The results of the questionnaire analysis proved that the awareness towards synchromodality was raised since 72% stated that after the playing the game they gained more awareness of the importance of flexibility while 48% agreed that they understood the importance of synchromodality. Interesting discussion between representatives from different sectors was generated and 64% agreed that they had enjoyed the session and the game.

Conclusions and Further Research

The main objective of this paper is to present a series of games that were developed to raise awareness towards synchromodality, a freight transport innovation expected to increase the efficiency of freight transport. Research shows that the horizontal and vertical collaboration between stakeholders is a prerequisite to achieve the synchronisation of the different modes and increase modal shift toward more environmentally friendly modes of transport (*1*). We used simulation gaming as our key research method to study the behavior and decision making of stakeholders of a complex system such as a transport network in a safe and controlled environment. Games offer participants an interactive, immersive, and motivating environment to express their attitudes and preferences. We have designed five games to help stakeholders understand the key challenges and opportunities offered by synchromodal transport. The results of the gaming sessions have shown a positive attitude change towards synchromodality after playing the games. The games we developed are both digital and board games. The principle advantages of the board game are the social cohesion and interaction offered by the board game environment. Participants have a shared view of the game environment and can interact naturally. The physical elements of the board game such as tokens, cards and pawns increase the fun element of an otherwise serious play. Board games also produce rich qualitative data. In combination with well-designed briefing and debriefing sessions board games can be very valuable research instruments even in today's digital era (*26*, *27*, *28*). A simulation metamodel has been developed using the design of the game where the choices of the players are inserted, and their effects are calculated. On the other hand, the digital games are a source of a lot of quantitative data that can be used to model the decisions of the players. Using the data from SynchroMania we try to use games as survey instruments and use them to model decisions. We have also developed simulation metamodels shown in Visser et al. (*24*) using the qualitative observations to create a computer simulation model of the board game to give quantifiable insights into the decision making of the stakeholders as if the game has been played few thousand times.

In future work we will try to organise more gaming sessions and collect more data to better understand the behavior of players. Through the application of games, answers to critical questions regarding synchromodality will be given. In addition, we have already started developing a game that investigates the preferences of shippers regarding the services and the costs of the synchromodal product. Finally, data will also be used to develop models shown in Visser et al. (*24*) that can be used to predict the future evolution of the behaviour of stakeholders.

Acknowledgments

The work presented in this paper is part of the SynchroGaming project funded by the Netherlands Organisation for Scientific Research (NWO) together with consortium partners comprising the rail, road and seaport authorities of the Netherlands.

Author Contributions

The authors confirm contribution to the paper as follows: study conception and design: Kourounioti, I.; Kurapati, S.; Lukosch, H.; Tavasszy, L. and Verbraeck, A; data collection: Kourounioti, I. and Kurapati, S.; analysis and interpretation of results: Kourounioti, I. and Kurapati, S.; draft manuscript preparation: Kourounioti, I.; Kurapati, S.; Lukosch, H.; Tavasszy, L. and Verbraeck, A. All authors reviewed the results and approved the final version of the manuscript.

References

1. Tavasszy, L., B. Behdani, and R. Konings. *Intermodality and Synchromodality*. https://papers.ssrn.com/sol3/papers.cfm?abstract_id=2592888. Accessed July 19, 2018.
2. Behdani, B., Y. Fan, B. Wiegmans, and R. Zuidwijk. Multimodal Schedule Design for Synchromodal Freight Systems. *European Journal of Transport and Infrastructure Research*, Vol. 3, No. 16, 2016, pp. 424–444.
3. Burnham, J. *Freigth Transport and the Modern Economy*. Routledge: Oxfordshire, 2014.
4. Schoemaker, T., K. Koolstra, and P. Bovy. Traffic in the 21st Centrury-A Scenario Analysis for the Traffic Market in 2030. In *The infrastructure Playing Field in 2030*, Delft University Press, Delft, 1999, pp. 175–194.
5. Zuidwijk, R., and A. Veenstra. The Value of Information in Container Transport. *Transportation Science*, Vol. 49, No. 3, 2015, pp. 675–685.
6. Tavasszy, L., and G. de Jong. *Modelling Freight Transport*. Elsevier, London, 2013.
7. Wolfe, J. Developing a Scientific Knowledge of Simulation/Gaming. *Simulation & Gaming*, Vol. 29, No. 1, 1998, pp. 7–19.

8. Greenblat, C. *Gaming-Simulation: Rationale, Design and Applications*. Sage Publications: New York, 1975.
9. Klabbers, J. Framework for Artifact Assessment and Theory Testing. *Simulation & Gaming*, Vol. 37, No. 2, 2006, pp. 155–173.
10. Duke, R. *Gaming: The Future's Language*. Beverly, CA, Sage Publications, 1975.
11. Meijer, S. *The Organisation of Transactions: Studying Supply Networks Using Gaming Simulation*. Wageningen University, Wageningen, 2009.
12. Bryman, A. *Business Research Methods*. New York: NY, Oxford University Press, 2015.
13. Voss, C. Case Research in Operations Management. *International Journal of Operations & Production Management*, Vol. 22, No. 2, 2002, pp. 195–219.
14. Popper, K. *The Logic of Scientific Discovery*. New York, Routledge, 2005.
15. Sommer, B., and R. Sommer. *A Practical Guide to Behavioral Research: Tools and Tehniques*. New York, Oxford University Press, 2002.
16. Klabbers, J. *The Magic Circle: Principles of Gaming & Simulation*. Rotterdam, Sense Publishers, 2009.
17. Kurapati, S., D. Groen, H. Lukosch, and A. Verbraeck. Microgames in Practice: A Case Study in Container Terminal Operations. In *ISAGA 2014: 45th Conference of the International Simulation and Gaming Association*, Dornbirn, Austria, 2015.
18. Kurapati, S., I. Kourounioti, H. Lukosch, G. Bekebrede, L. Tavasszy, A. Verbraeck, D. Groen, J. van Meijeren, and L. Lebesque. The Role of Situation Awareness in Synchromodal Corridor Management: A Simulation Gaming Perspective. In *20th EURO Working Group on Transportation Meeting, EWGT 2017*, Budapest, Hungary, 4–6 September 2017.
19. Biedel, J. State of the Art of Serious Games for Business and Industry. In *17th International Conference on Concurrent Enterprising*, Aachen, 2011.
20. Synchromodaliteit. Synchromodaliteit, 2017. http://www.synchromodaliteit.nl/. Accessed July 19, 2017.
21. Kurapati, S., I. Kourounioti, H. Lukosch, L. Tavaszzy, A. Verbraeck, L. van Veen, and B. van Nuland. *Exploring Challenges and Solutions for Container Transportation Using Rail - A Modeling and Simulation Game Study*. Naples, Italy, 2017, IEEE.
22. Kurapati, S., M. Freese, I. Kourounioti, H. Lukosch, G. Bekebrede, T. Smit, J. van Meijeren, B. van Nuland, and L. van Veen. Attitude Measurement with Board Games in Transportation Modes. In *ISAGA Conference*, Deft, Netherlands, 2017.
23. Lukosch, H., S. Kurapati, D. Groen, and A. Verbraeck. Microgames for Situated Learning: A Case Study in Interdependent Planning. *Simulation & Gaming*, Vol. 47, No. 3, 2016, pp. 346–367, 2016.
24. Visser, J., R. Konnings, and B. Pielage. A New Hinterland Transport Concept for Rotterdam: Organisational and/or Technological Challenges. In *Transportation Research Forum*, Upper Great Parts Transportation Institute, North Dakota State University, North Dakota, 2007.
25. Lebesque, L. *TNO Time*, TNO, 2015. https://time.tno.nl/en/articles/synchromania-teaches-students-the-art-of-synchromodal-thinking/. Accessed July 20, 2017.
26. Kourounioti, I., and L. Tavasszy. Can We Apply Serious Gaming as An Alternative to Stated Preference Experiments? In *International Choice Modeling Conference*, Cape Town, South Africa, 2017.
27. Lukosch, L., S. Kurapati, G. Bekerbrede, S. Tiemersma, L. van Veen, and A. Verbraeck. Design Considerations for Building a Scalable Digital Version of a Multi-player Educational Board Game foa MOOC in Logistics and Transportation. In *Games and Learning Alliance Conference (GALA)*, 2016, pp. 167–176.
28. Meijer, S. The Power of Sponges: Comparing High-tech and Low-tech Gaming for Innovation. *Simulation & Gaming*, Vol. 46, No. 5, 2015, pp. 515–523, 2015.

The Standing Committee on Visualization in Transportation (ABJ95) peer-reviewed this paper (18-04269).

TRR
JOURNAL OF THE TRANSPORTATION RESEARCH BOARD

Article

Transportation Research Record
2018, Vol. 2672(44) 82–92

DOI: 10.1177/0361198118778930
journals.sagepub.com/home/trr
SAGE

The CLIOSjre Process: A Flexible Multicriteria, Multistakeholder Decision Framework for Transportation Planning under Uncertainty

Joanna Moody[1], Timothy Patton Doyle[1], Scott Middleton[1], and Joseph Sussman[1]

Abstract
Transportation investments determine the long-term success or failure of a transportation provider. It is therefore vital for decision makers to have an in-depth understanding of the alternatives available before they choose to invest. However, often, the process of evaluating alternatives is lengthy, costly, and contentious, particularly for transportation infrastructure investment decisions that are large, complex, and interconnected with other economic development and sustainability goals. Furthermore, transportation investments involve many decision makers, each with different priorities and expertise. Therefore, there is a need for transparent, accurate, flexible, and practicable decision-making aids that can handle the complex challenges facing the decision-making bodies of transportation providers and planning organizations. This paper introduces a new decision aid—the CLIOSjre Process—that combines insights from multicriteria decision analysis, multistakeholder negotiation theory, and uncertainty analysis. The CLIOSjre Process helps decision makers compare multiple alternatives across multiple objectives and seek an informed collective transportation investment decision. Unlike other multicriteria decision aids, the CLIOSjre Process accounts for differences of opinion among decision makers and is designed to facilitate constructive negotiation among them. Finally, the CLIOSjre Process formally accounts for sources of uncertainty inherent in these decisions. In this way, the CLIOSjre Process provides a unique and flexible framework for investment analysis that can adapt to changes in transportation alternatives, decision-maker priorities, and emerging real-world conditions. The usefulness of this new decision aid is illustrated for the East Japan Railway Company's consideration of a transportation investment opportunity in high-speed rail development on the Northeast Corridor of the United States.

In both the public and private sector, investment decisions often determine the long-term success or failure of an organization. This is especially true for transportation providers who rely heavily on their built infrastructure. Transportation providers that invest wisely in infrastructure often experience increasing ridership, increasing revenue, and a continually improving level of service for their customers. Organizations that do not make appropriate investments can experience declining ridership, revenue, and level of service in the long term (*1*).

Due to the importance of these investment decisions, it is vital for all decision makers to have an in-depth understanding of the alternatives available before they choose to invest (*1*). But often this alternatives evaluation process is lengthy, costly, contentious, and subject to the pressures of political economy. This is particularly true for transportation infrastructure investment decisions because of their scale, complexity, highly technical details, and interconnectivity with other economic development and sustainability goals. Furthermore, transportation investments involve many decision makers, each with different types of expertise and priorities. Therefore, there is a need for transparent, accurate, flexible, and useable decision-making aids that can handle the complex challenges facing the decision-making bodies of transportation providers and planning organizations.

In addition, the possibility of rapid changes in transportation technology and travel patterns necessitates

[1]Department of Civil & Environmental Engineering, Massachusetts Institute of Technology, Cambridge, MA

Corresponding Author:
Address correspondence to Joanna Moody: jcmoody@mit.edu

frameworks that adapt to organization change, reflect evolving priorities, and expedite high-level decision making. Transportation investment decisions are time-sensitive (i.e., there is an opportunity cost of delay). Any decision-making aid must deliver a robust, defensible, and understandable result quickly and should be modular to incorporate changes in realities or perspectives.

This paper builds upon previous research into the CLIOS Process that represents transportation systems as CLIOS Systems that are composed of multiple interrelated components and subsystems (Complex); have impacts that are large in magnitude, are long-lived, and have a broad geographical extent (Large-scale); are often interconnected with other sociotechnical systems, such as the environment or economy (Interconnected); explicitly include social, political, and economic aspects beyond the technical or engineered system (Open); and have physical components and subsystems are connected directly to the social context within which they operate (Sociotechnical) (*2*). Drawing upon the CLIOS Process, the existing state of practice in multicriteria decision analysis, and principles of multistakeholder negotiation, the CLIOSjre Process was developed as a new decision aid that:

1. Takes into consideration all of the potential impacts of the transportation investment decision by evaluating it using multiple criteria, synthesizing both qualitative and quantitative information.
2. Accounts for the opinions of multiple decision makers and provides equal information to all involved in making an informed judgment. The CLIOSjre Process is designed to facilitate negotiation between multiple decision makers who may have very different levels of technical knowledge and backgrounds.
3. Identifies and addresses the two primary sources of uncertainty in analysis: measurement (metric) error and real-world uncertainty.

This CLIOSjre Process was developed to understand investment choices for the East Japan Railway Company (JR East) in expanding into international high-speed rail (HSR) development on the Northeast Corridor (NEC) of the United States. This case study is used to illustrate the steps of the CLIOSjre Process and to draw conclusions about its usefulness in expediting the process of transportation planning and investment decisions that consider multiple objectives and the needs of multiple stakeholders under significant uncertainty.

The Case Study: JR East and International HSR Development

JR East has saturated much of their domestic HSR market and established their Shinkansen brand as one of the safest, most efficient, and most reliable systems in the world. Thus, the company is uniquely positioned to expand into international HSR development. However, there are many potential international HSR markets, and within each market there are many interrelated factors and possible business opportunities for the company to consider. To expand successfully overseas, JR East must carefully select which market(s) to pursue and the best way to proceed in those markets. Such choices require a decision-making aid that can evaluate multiple investment alternatives across different criteria.

Furthermore, JR East's senior-level management has limited experience in international HSR markets and is still developing the company's business strategy for these markets. The decision-making body is made up of individuals with distinct and sometimes conflicting perspectives with different backgrounds and levels of technical expertise. Therefore, any decision-making aid will need to account for multiple decision-maker perspectives and transparently communicate tradeoffs among priorities in a way that all decision makers can understand.

Finally, any decision to invest in international HSR markets involves significant uncertainty, both in terms of the accuracy of available data and analytical models (measurement error) and in terms of how the system will develop in the future (macro uncertainty). Therefore, the decision-making aid will need to consider multiple sources of uncertainty and succinctly communicate the relative risks involved in different investments.

In response to these needs, MIT's Regional Transportation Planning and High-Speed Rail Research Group developed the CLIOSjre Process to help JR East consider which international HSR markets it should pursue and how it might pursue them. This paper presents the CLIOSjre Process and its application to JR East's potential business opportunities in the HSR development of the NEC, but this framework is generally applicable to other decisions within transportation investment and planning that require robust, transparent, and adaptive decision making in the context of multiple criteria, multiple stakeholder priorities, and significant uncertainty. Though the generalizability of the framework to other applications and other stakeholder perspectives (particularly in the public rather than private sector) is an area of future research, this paper explains each step of the CLIOSjre Process in general terms, illustrates each step with the specific case study of JR East and the NEC, and then points to how each step might be modified to fit other contexts.

The CLIOSjre Process

In the CLIOSjre Process, JR East or another decision maker is faced with a choice between one of many

Table 1. Example CLIOS Bundles for High-Speed Rail Development on the Northeast Corridor

#	Physical infrastructure design	Organizational structure	Funding mechanism
1	Incremental HSR – existing alignment – shared track	Vertically integrated – Amtrak	Public infrastructure – public operations
2	International quality HSR (Boston to New York City); incremental HSR (New York City to Washington, D.C.) – new alignment – dedicated track	Vertically separated – non-Amtrak single operator	Public infrastructure – private operations
3	International quality HSR (Boston to D.C.) – new alignment – dedicated track	Vertically separated – competing operators	PPP infrastructure – private operations

Note: HSR = high-speed rail; PPP = public-private partnership.

investment alternatives, or CLIOSjre bundles. These bundles are packages of investment alternatives (or an investment program) and policies that describe a particular development path for a transportation system plus a business role that the decision maker can take in that system. In order to understand their merits, each bundle is evaluated using a number of metrics (value criteria). For each of the metrics, each bundle is assigned a (metric) grade (from A to F). Informed by the decision-maker's preferences, a set of weights (referred to collectively as a strategy vector) combines the individual metric grades into a single overall grade for each bundle. By comparing these overall grades as well as the metric grades, decision makers can arrive at a better understanding of the tradeoffs among various investment alternatives and can negotiate a more informed decision.

The CLIOSjre Process described above comprises eight steps, each of which is explained in more detail in the following subsections:

1. Gather Inputs
2. Identify CLIOSjre Bundles
3. Determine Metric Grades
4. Check CLIOSjre Bundle Acceptability
5. Choose Strategy Vector(s)
6. Calculate Overall Grade
7. Consider Uncertainties
8. Communicate Results to Decision Makers and Faciliate Negotiation

Gather Inputs

The CLIOSjre Process takes as its inputs a set of transportation investment alternatives. In its application to the NEC, the CLIOSjre Process was designed to follow a CLIOS Process study, which identified and evaluated CLIOS bundles of strategic alternatives from the perspective of the HSR system (not from the perspective of JR East) (*2*). Developed and applied to a diverse range of applications over the past two decades at MIT, the CLIOS Process explores how the physical infrastructure and institutional context of an HSR system might develop in the future and highlights key uncertainties. The CLIOS Process provides a flexible framework in which various analytical methods – such as benefit–cost analysis, financial analysis, and scenario planning – can be applied to demonstrate the value and trade-offs among different strategic alternatives for HSR development and provide in-depth knowledge of the needs of the HSR system (*2*).

For the case study of the NEC, the CLIOS Process identified and evaluated bundles of strategic alternatives that considered the physical infrastructure design, organizational structure, and funding mechanisms for the HSR development (see Table 1). These CLIOS bundles for the NEC are adapted from an earlier study, which includes detailed explanations of each strategic alternative for the system development (*3*) and were updated based on the results of the NEC Future Tier 1 Environmental Impact Statement (*4*).

Although the CLIOSjre Process was designed to follow the CLIOS Process, other sources of information could be used to identify the investment alternatives under consideration for the system. Alternative sources of input might include project plans, feasibility studies, environmental and social assessments, and so forth. Therefore, the CLIOSjre Process is a stand-alone decision-making aid that can be modified to evaluate any investment program (not just bundles from the CLIOS Process) from the perspective of any decision maker.

Identify CLIOSjre Bundles

The CLIOS bundles of strategic alternatives must be modified with additional information so that the CLIOSjre Process can evaluate the HSR development from the perspective of the decision maker (JR East) rather than the stakeholders of the system. In particular, the analyst must identify the possible business roles that the decision makers may take and combine them with the CLIOS bundles:

$$\text{CLIOS Bundle} + \text{Business Role} = \text{CLIOSjre Bundle}$$

Just as the investment alternatives considered within the CLIOS bundles will depend on the transportation system under consideration, the business roles available to the transportation provider will depend on the particular characteristics of the organization and the market. Decision-maker input is necessary to accurately determine which business roles are viable given financial, human resources, and other constraints. In this case, collaborative discussion with representatives from JR East determined that the company could enter international HSR markets in multiple business roles, including (but not limited to):

1. Do not become involved in this HSR market.
2. Provide planning, engineering, or operations consultation (e.g., demand analysis, track design, or operations evaluation).
3. Operate a system under concession, including maintenance.
4. Operate a system by leasing track capacity from a separate infrastructure manager.

If applying the CLIOSjre Process for a public transportation agency rather than a private sector provider, public roles could be substituted for business roles. These public roles may differ depending on, for example, whether the public agency procures the investment program using a traditional design-bid-build delivery method, engages a construction manager, or pursues a public–private partnership arrangement.

These business (or public) roles are combined with CLIOS bundles (or investment programs) to form CLIOSjre bundles (see Table 2). Not all combinations may be logically feasible given the relationship between the business role and the strategic alternatives within the CLIOS bundle. Therefore, the analyst must review all possible CLIOSjre bundles and eliminate infeasible ones. For example, a business role of JR East operating HSR service by leasing track capacity is logically incompatible with a CLIOS bundle that has Amtrak as the single operator (see Table 2, number 1.4).

Once infeasible CLIOSjre bundles are removed, the analyst selects which CLIOSjre bundles to analyze further. The selection of CLIOSjre bundles is often driven by professional judgment; however, using experimental and statistical principles such as factorial design, the selection of bundles can be optimized to allow the analyst to draw conclusions about the performance of other feasible CLIOSjre bundles not directly analyzed (*5*).

Determine Metric Grades

The CLIOSjre bundles are evaluated according to metrics, which represent the decision maker's different objectives or priorities in the investment decision. These metrics should reflect characteristics of the project and characteristics of the decision-making organization itself. The metrics should be independent to improve the accuracy of the decision model and to avoid double counting. Furthermore, the metrics must be designed to be inclusive of all major decision-maker objectives and applicable across all CLIOSjre bundles for the investment.

For the NEC, 10 metrics were identified that address JR East's priorities in their international HSR business strategy:

M1. Expansion potential (to other international HSR applications),
M2. Expected profit
M3. Competition
M4. Cooperation
M5. Flexibility
M6. Public benefits
M7. Environmental impacts
M8. Strengths and weaknesses
M9. Reputation for excellent service
M10. Human resource development.

The choice of metrics included in the analysis will depend on the context of the CLIOSjre Process application, including the market or transportation system and the stakeholder perspective being considered. Therefore, although some of the 10 metrics presented here, such as M6. Public Benefits and M7. Environmental Impacts, may be considered for almost all transportation systems and stakeholder perspectives (whether the decision makers are public or private), others may be particular to the

Table 2. Matrix of CLIOSjre Bundles

	Business role			
CLIOS bundle	1. No involvement	2. Consulting	3. Operation concession	4. Operate by leasing track capacity
1	*1.1*	*1.2*	1.3	1.4
2	**2.1**	*2.2*	2.3	2.4
3	**3.1**	**3.2**	*3.3*	*3.4*

Note: Bold indicates those selected for analysis, italics indicates those that are feasible but not selected for analysis, and underlined indicates those that are strictly infeasible.

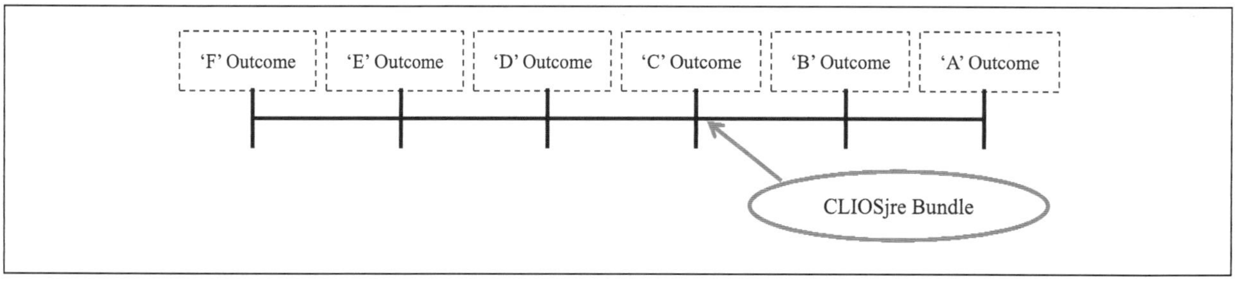

Figure 1. Placing a CLIOSjre bundle on the representative grade spectrum.

case of NEC HSR or the interests of JR East, or to both (such as M10. Human Resource Development).

Once the metrics are determined, the next step of the CLIOSjre Process is to grade each CLIOSjre bundle according to each metric. Although this step sounds straightforward, the synthesis of both qualitative information and quantitative information into grades for each metric often poses a significant challenge. The CLIOSjre Process uses a generic method to grade all metrics. This method must provide a defensible combination of the different types of information while remaining transparent to decision makers who must understand the outputs of the analysis. This step in the CLIOSjre Process relies on the capacity and professional judgment of the analyst. Therefore, it is critical that all assumptions, methods, and conclusions are synthesized in a way that is accessible for decision makers.

Generic Metric Analysis Method. In order to map each metric evaluation result to a letter grade, the CLIOSjre Process uses a spectrum of representative outcomes (*6*, *7*). First, the analyst develops representative outcomes for each metric ranging from ideal (A) to failing (F) (see Figure 1). Once this spectrum has been established, the analyst compares the result of the metric evaluation for each of the CLIOSjre bundles to the representative outcomes along the grade spectrum. Based on its location along the spectrum, each CLIOSjre bundle receives a grade for that metric (see Figure 1). This process is repeated for all CLIOSjre bundles and all metrics.

The CLIOSjre Process's use of a grade spectrum to synthesize both quantitative and qualitative information from different metric analyses provides a common multicriteria framework that is flexible to the introduction of new objectives (metrics) or input data. By mapping each metric analysis to a representative grade spectrum, analysts can express the results of different analytical methods on a common measurement scale that is familiar to all decision makers. This allows decision makers to compare the performance of a CLIOSjre bundle across different metrics.

Specific Metric Analysis Method. The representative outcomes used to establish the grade spectrum for each metric are based on different evaluation techniques and performance measures. For some metrics (such as M2. Expected Profit) existing techniques (like financial analysis) may be used to assign a grade. For other metrics, existing methods may need to be adapted or new frameworks established to evaluate the CLIOSjre bundle. As an illustrative example, Table 3 provides a brief explanation of the 10 metric analysis methods used to evaluate the CLIOSjre bundles for the NEC case study.

The choice of metrics and analysis methods will depend on the transportation investment program or the stakeholder perspective, or both, being considered in the particular application of the CLIOSjre Process. For example, in the case of a public sector transportation agency, a metric that considers the affordability of the project to the public sector (considering operating revenue and long-term subsidy needs) may be substituted for M2. Expected Profit.

Check CLIOSjre Bundle Acceptability. Once a CLIOSjre bundle has received a grade for each metric, the analyst determines whether the CLIOSjre bundle is acceptable to the decision maker based on a predetermined minimum grade for each metric. CLIOSjre bundles that do not meet the necessary minimum grades are discarded without further analysis. For example, in the NEC case study JR East might decide not to pursue any CLIOSjre bundle that receives a grade of D or lower for M2. Expected Profit. The objective of this step is to speed up the process by reducing unnecessary analysis in the evaluation of the market and to eliminate non-critical information or irrelevant alternatives from the discussion of tradeoffs.

Choose Strategy Vector(s). For those CLIOSjre bundles not discarded, the analyst combines metric grades into an overall grade. Using the 10 metric grades for each NEC CLIOSjre bundle as inputs, the CLIOSjre Process applies a set of strategy vectors, each consisting of 10 weights $\{w_1, .., w_{10}\}$, to these individual grades to compute an overall

Table 3. Metric Analysis Methods for the 10 Metrics Used to Evaluate JR East's Investment Opportunities for HSR Development in the NEC

Metric	Method
M1. Expansion potential	Opportunities analysis – qualitative mapping of the CLIOSjre bundle with the number and attractiveness of any relevant business investment opportunities that the investment being considered might open up.
M2. Expected profit	Financial analysis, or industry benchmarking – quantitative determination of the financial return on investment (ROI) of the CLIOSjre bundle. The higher the ROI, the higher the grade for M2.
M3. Competition	CLIOSjre six forces analysis – an adaptation of Porter's five forces analysis (*8*) that qualitatively determines how well the CLIOSjre bundle aligns with the six forces of market competitors, business role competitors, market substitutes, business role substitutes, power of suppliers, and power of buyers. The fewer of the six competitive forces faced in the CLIOSjre bundle, the higher the grade for M3.
M4. Cooperation	Predictive coalition building analysis (PCBA) – a method that assigns a grade to the CLIOSjre bundle based on the number of potential partners with similar interests among the stakeholders of the system *and* the degree of incentive that these partners have for working with the decision maker (*9*).
M5. Flexibility	Flexibility analysis – a qualitative and quantitative adaptation of real options analysis (*10*) focused on evaluation of business risk and flexibility within the CLIOSjre bundle. The CLIOSjre bundle is evaluated based on whether it has the inherent option to abandon, expand, or delay the investment under different future scenarios.
M6. Public benefits	Benefit–cost analysis (BCA) – quantitative determination of the economic internal rate of return (EIRR) of the project, with the key monetized social benefits being travel time savings and lives saved. The higher the EIRR of the project, the higher the grade for M6.
M7. Environmental impact	Environmental impact analysis – qualitative analysis of the number of impacts to existing natural resources and environmental justice communities, prudent mitigation measures, and feasibility of alternatives (as defined in the National Environmental Policy Act of 1970). Grades are assigned based on the severity of impacts for the CLIOSjre bundle and the number of feasible and prudent alternatives.
M8. Strengths and weaknesses	Strengths and weakness analysis – a partial SWOT (strengths, weaknesses, opportunities, and threats) analysis (*11*). The greater the number of strengths and the fewer the weaknesses that match the CLIOSjre bundle, the higher the grade for M8.
M9. Reputation for excellent service	Brand impact analysis – a methodology similar to Environmental impact analysis (M7) that identifies the reputation impacts on the service speed, safety, reliability, hospitality, and environment of the CLIOSjre bundle and the number of feasible and prudent alternatives.
M10. Human resources development	Attract, train, and retain analysis – qualitative determination of the CLIOSjre bundle with the number of opportunities it has for attracting, training, and retaining capable employees.

grade for the bundle according to a basic linear combination (just as one would calculate a weighted grade point average [GPA] on a standard 4.0 scale):

$$Overall\,Grade = M1\,Grade\ x\ w_1 + M2\,Grade\ x\ w_2 + \ldots + M10\,Grade\ x\ w_{10}$$

The choice of strategy vector has a large influence on the overall grade of each CLIOSjre bundle. If decision makers can all agree on a single strategy vector that is representative of the organization's priorities, then one overall grade can be calculated for each CLIOSjre bundle. However, one challenge inherent in decision making in transportation organizations is the possibility of differences of opinion about the priorities of the company and, therefore, what weights to choose to represent those interests. One way to resolve this problem is to hold a transparent conversation about strategy vectors, weights, and overall grades in a particular market. In many cases, it is helpful to come up with several example strategy vectors that represent possible conflicting business priorities for the decision maker. For example, in the NEC case, strategy vectors were selected that emphasized expected profit, international brand and service reputation, or flexibility to risk and uncertainty. Then overall grades are calculated for the different strategy vectors to show the sensitivity of the bundle's overall evaluation to the relative priorities of the decision makers (see Figure 2).

Calculate Overall Grade

Once the strategy vector(s) have been selected, the analyst calculates the overall grade using the additive structure presented above. The CLIOSjre Process makes use of this additive approach because it:

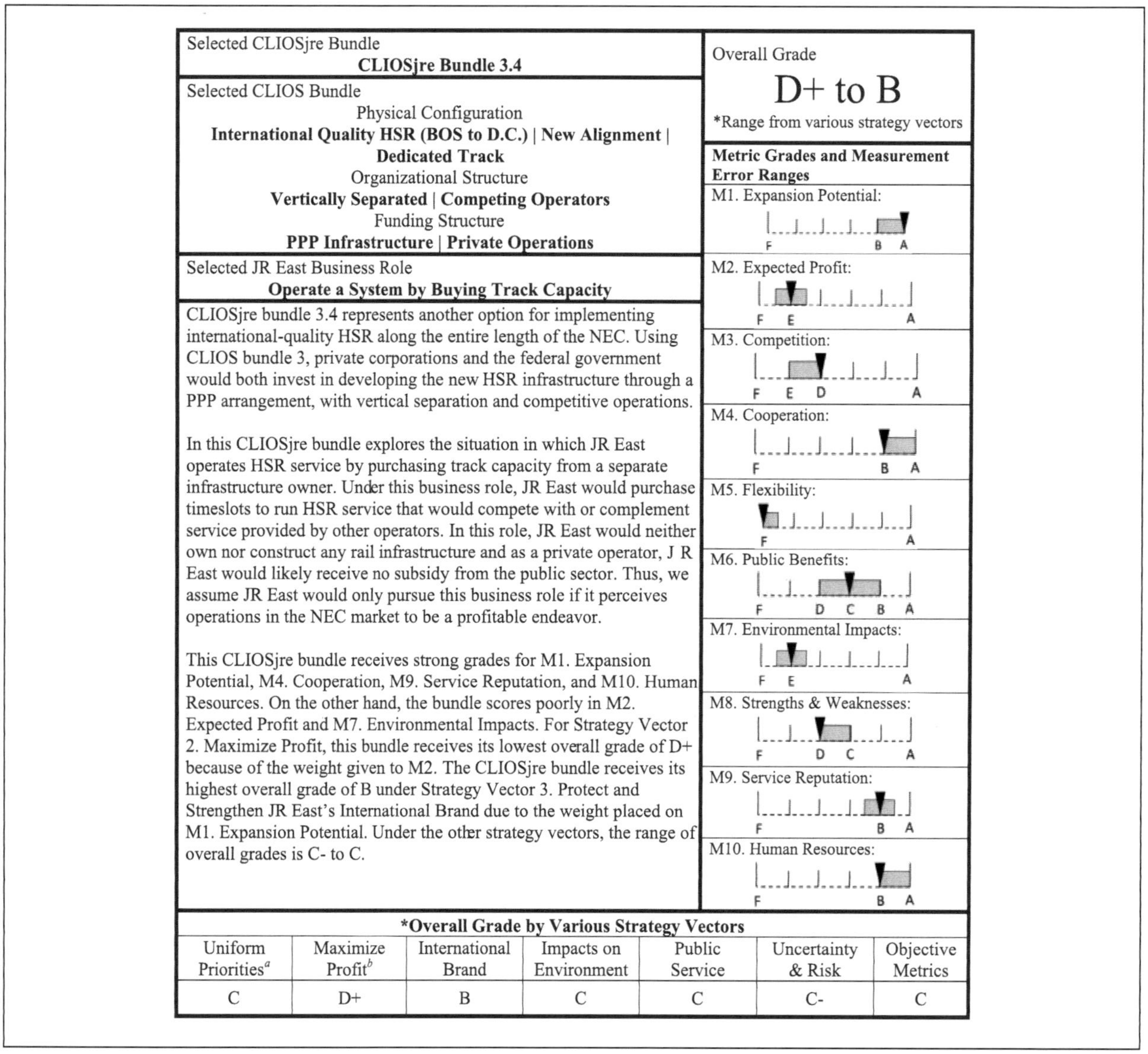

Figure 2. Example CLIOSjre assessment.

[a]This reference strategy vector assigns equal weights of 0.10 for each of the 10 metrics.

[b]As an example of non-uniform priorities, this strategy vector assigns a weight of 0.60 to M2. Expected Profit, a weight of 0.10 to M1. Expansion Potential, a weight of 0.09 to M8. Strengths and Weaknesses, and a weight of 0.03 to each of M3, M4, M5, M6, M7, M9, and M10.

1. Is intuitive, improving the transparency of the CLIOSjre Process;
2. Facilitates the selection of weights as the relative importance of metrics is easily observed; and
3. Is widely used in the multicriteria decision-making literature (*12*).

For any market, it is informative to consider how the CLIOSjre bundles compare with each other according to various strategy vectors. Framing the CLIOSjre Process results by different strategy vectors allows analysts and decision makers to visualize and discuss the relative ranking of different CLIOSjre bundles from the perspective of others, who may have different priorities among the many metrics.

Consider Uncertainties

Understanding uncertainty in the overall grade of each CLIOSjre bundle is an important part of the CLIOSjre Process. There are two main sources of uncertainty in the CLIOSjre Process, which are treated separately:

1. Measurement uncertainty in the determination of metric grades.
2. Uncertainty in the strategy vector used to calculate the overall grade.

Measurement uncertainty in each of the metric grades results from the imperfections of the analytical framework and performance measure in accurately capturing the "true" current conditions. To estimate the measurement error for each CLIOSjre metric, the CLIOSjre analyst performs a sensitivity analysis for each metric. Although the form of this sensitivity analysis is different for each metric, the objective is the same: to identify the range of reasonable metric grades for each bundle given different sources of uncertainty. These sensitivity analyses vary the assumptions for each metric but hold the analysis methodology constant. There is no guarantee that the true metric grade lies within the measurement error for each metric, but the analyst can be confident that the measurement error includes the range of metric grades that a reasonable analyst would find given the information that exists today.

The other source of uncertainty is rooted in the weight applied to each metric grade. By using several strategy vectors to combine the metric grades, the decision makers can learn more about how the choice of strategy vector affects the overall grades of the bundles.

Communicate Results to Decision Makers and Facilitate Negotiation

The final step is to compile and communicate the results from the individual metric analyses and calculation of overall grades for each of the CLIOSjre bundles under consideration. The analyst should discuss any assumptions inherent in the analysis and present key tradeoffs among different CLIOSjre bundles according to metrics and strategy vectors. These tradeoffs must be communicated in a way that is easy to understand and facilitates discussion among decision makers with different priorities and levels of expertise. Based on insights from multistakeholder negotiation theory, the most effective communication of results will (*13*):

- Focus the decision makers on cooperative problem solving rather than competitive negotiation;
- Identify opportunities for the participants rather than limitations; encourage decision makers to think outside the box;
- Encourage participants to consider the perspective of other participants so they can identify solutions that are beneficial to the other decision makers; and
- Identify and explain all types of uncertainty so that key decision makers understand the risks that are present.

For each CLIOSjre bundle evaluated for the NEC, the analyst creates an assessment sheet that presents the metric grades (and their error ranges) and comments on the distribution of overall grades by strategy vector. These assessment sheets present the results of the CLIOSjre analysis in a succinct and standardized format.

For these assessment sheets, as with other dashboard-type presentations of synthesized results, there is a tradeoff between content and legibility. If too little information is displayed on the assessment sheet, it is not particularly useful. On the other hand, too much information makes it difficult to identify the most important information and may frustrate the decision maker. In addition to presenting an appropriate level of information, the assessment sheets should be formatted for maximum legibility. Well-formatted assessments allow the decision maker to quickly identify and understand the information relevant to the decision. Although there are many suggestions of how to format technical information (*14*), there are few well-defined standards.

The research team iterated through multiple versions of these assessment sheets, testing them with representatives from JR East, to determine how much information to present and how to format it (see Figure 2). Rather than present a single overall grade for a given CLIOSjre bundle, the assessment sheet provides a range of overall grades (top left) based on the choice of strategy vector (listed at the bottom). These strategy vectors represent the priorities of different individuals on the Executive Board of JR East, so the range of overall grades for a given bundle illustrates how the investment decision on a single CLIOSjre bundle might vary based on the evaluations of different decision makers. By not choosing a single strategy vector to calculate one overall grade, the CLIOSjre Process focuses decision makers on tradeoffs that foster cooperative (rather than competitive) negotiation and encourage decision makers to consider the perspectives of others. To highlight the impacts of measurement error and macro uncertainty, each individual metric grade is expressed as a range (left-hand column). This explicitly synthesizes the risks that may be present in different dimensions of the decision.

Although the CLIOSjre assessment follows key principles of constructive multistakeholder negotiation and represents the needs of an actual, diverse decision-making body (JR East), the form of the assessment sheet in Figure 2 may not be the optimal way to display the results for any audience. Depending on how these CLIOSjre assessment sheets are used in practice, the formatting of these assessments may need further refinement, especially if a new case study has a significantly different number of metrics or strategy vectors.

Conclusions and Future Research

As organizations must adapt and innovate in the face of changing mobility patterns, the introduction of new technologies, and the increased focus on integrated transportation planning, their decision-making tools must be flexible and responsive to changes and uncertainties. The CLIOSjre Process was developed as a multicriteria, multistakeholder decision aid, which addresses uncertainty and tradeoffs among multiple objectives in a transparent and flexible framework. In particular the CLIOSjre Process:

1. Evaluates transportation investments using multiple criteria, which can capture the impact of the alternative on interconnected systems such as the environment, economy, and urban form;
2. Takes into consideration all objectives of decision makers even when these objectives are in conflict with one another. The CLIOSjre Process is designed to facilitate negotiation between multiple decision makers who may have very different levels of technical knowledge and backgrounds; and
3. Identifies and addresses the two primary sources of uncertainty in analysis.

Although traditional investment evaluation tools are occasionally adapted to include one or two of these features, the CLIOSjre Process is able to address all three. The CLIOSjre Process provides a more in-depth and robust understanding of a transportation investment or planning decision, while remaining adaptive to new decision-maker priorities, new possible development alternatives, and new sources of data.

This paper demonstrated the usability of the CLIOSjre Process in a case study of JR East's role in international HSR development on the NEC of the United States. Representatives from JR East were involved in the definition of metrics, business roles, and strategy vectors that represented the priorities, capabilities, and business interests of the organization. Throughout the 3-year development, this list was updated to reflect changes in management and business priorities, testing the adaptability of the method to organizational change. Recognizing the value of this process beyond any single investment decision in the NEC, JR East is now internalizing this process for decision making in international HSR markets and exploring how it may be modified to inform other aspects of their business.

Value of the CLIOSjre Process

Using insights from multicriteria decision analysis, multistakeholder negotiation theory, and uncertainty theory, the CLIOSjre Process is designed to help decision makers compare multiple alternatives and make an informed transportation investment decision.

The CLIOSjre Process evaluates each investment alternative using multiple criteria, effectively capturing the multifaceted impacts of transportation decisions on the economy, society, the environment, and the financial viability of the operator. By considering each metric separately, the CLIOSjre Process provides a detailed understanding of each alternative. The CLIOSjre Process then uses a linear weighting scheme to combine the CLIOSjre metric evaluations into a single overall grade for each alternative. This overall grade provides the decision maker with an actionable ranking of the alternatives. Thus, the CLIOSjre Process provides both a detailed and holistic understanding of the alternatives available to the decision maker.

Transportation investment decisions often involve multiple decision makers with different priorities. By applying multiple strategy vectors to combine the metric grades, decision makers can consider how investment decisions may be affected by the priorities of their peers. By presenting outputs from multiple perspectives within a single decision-making body, the CLIOSjre Process facilitates negotiation among decision makers; even if they have competing objectives, the transparency of CLIOSjre Process is designed to help decision makers understand key tradeoffs and find common ground.

In addition to addressing the concerns of multiple decision makers, the CLIOSjre Process formally identifies and addresses uncertainty in the analysis. Uncertainty in long-term transportation investments is a significant source of risk. Thus, by working to understand this uncertainty, the CLIOSjre Process enables stakeholders to put their trust in the outcomes of the analysis.

Therefore, the CLIOSjre Process is a flexible multicriteria decision aid that is designed to facilitate constructive negotiation by acknowledging differences of opinion among decision makers; to formally account for sources of uncertainty; and to adapt to changes in technological and organizational strategic investment alternatives, decision-maker priorities, and emerging real-world conditions. Though designed for transportation systems, these features make the CLIOSjre Process applicable to other decisions that involve risky, contentious, and long-lived outlays of capital on CLIOS Systems. Despite these advantages, the CLIOSjre Process may not be suited for all situations. The next sections identify the limitations of the CLIOSjre Process and propose future research to improve the utility of the CLIOSjre Process.

Limitations of the CLIOSjre Processes

Quality and Availability of Data and Analytical Methods. As with any other analytical framework or decision-making aid, the CLIOSjre Process relies on various sources of input data and analytical methods to evaluate the investment alternatives according to the individual metrics. Therefore, the results obtained from the CLIOSjre Process are only as good as the quality of available data and analytical techniques applied within it. This again highlights the importance of considering metric error in addition to more macro-level sources of uncertainty.

Combination of Independent Metrics. In the CLIOSjre Process, the analyst combines the independent CLIOSjre metrics into a single overall grade for each alternative. This overall grade allows the decision maker(s) to easily understand and compare the results for each alternative. However, using a single grade can imply false equivalencies or obscure the inherent complexity of the tradeoffs between alternatives (*12*). Therefore, the CLIOSjre Process may sacrifice representativeness of the true complexity of the decision in order to create an approachable and practicable result.

Linear Weighting Scheme. The linear weighting scheme used in the CLIOSjre Process may not be as useful as other methods of aggregating multiple criteria. By assuming a linear weighting scheme, the CLIOSjre Process is restricted in its ability to represent more complex preferences (*12*). Thus, although the linear weighting scheme is intuitive and easy to understand, this linear weighting scheme reduces the subtlety of the CLIOSjre Process in representing the decision makers' preferences.

Negotiating with Fundamental Disagreement. Although the CLIOSjre Process can aid the decision-making process by allowing for negotiation amongst decision makers, it cannot guarantee a group decision. Each decision maker has their own definition of value for the system, and these value differences can result in stalemate when logic would otherwise dictate that there is room for agreement (*13*). A disagreement on fundamental values can make it difficult to agree upon a common set of metrics to evaluate the system and the relative importance of those metrics embodied in the strategy vector. Thus, when decision makers disagree on fundamental objectives, the CLIOSjre Process provides a common output for discussion but does not guarantee a single outcome to the negotiation.

Future Research

Although the CLIOSjre Process is ready to be applied to problems in both the public and private sector, there are a number of methodological extensions that may improve the usefulness of the process to decision makers and increase its range of application.

Combination of Multiple Metrics. For the CLIOSjre Process, a linear weighting scheme is used to combine the independent CLIOSjre metrics into an overall grade for each alternative. Although this linear scheme has several advantages, it cannot always reflect the decision makers' preferences. There are other methods for combining independent metrics identified in the literature (e.g., outranking methods, utility models, non-classical approaches). It is possible that one or more of these other methods is equally transparent, flexible, and usable while also more accurate in its representation of the decision maker's preferences.

Soliciting Preference from Decision Makers. The CLIOSjre Process determines the weights for the CLIOSjre metrics from discussion with decision makers. Although this direct elicitation of weights from decision makers is transparent, it may not be the most effective way to determine their preferences. When directly eliciting linear weights for combining metrics, the decision maker may be able to assign relative importance of various metrics, but the particular values of the weights may be arbitrary (*12*).

One alternative to direct solicitation is determining the weights through a series of questions framed in terms that mean something to the decision maker. The decision-maker's choice would indirectly imply the relative value of a pair of metrics and their appropriate relative weights. Although the CLIOSjre Process is designed for a direct solicitation of weights, the process is compatible with an indirect solicitation process. Further research could determine if an indirect solicitation technique is a more effective method for soliciting preferences for the CLIOSjre Process (*1*).

Translating the CLIOSjre Process from the Private to the Public Sector. Though the CLIOSjre Process is flexible and has potential for applications in other decision-making contexts in which the possible solution sets are complex and potentially controversial, it has only been tested in a single case based on a private company's decision. Applying this process from a public sector stakeholder's perspective may involve additional challenges and risks that would need to be considered and mitigated. The process requires direct, substantive, and iterative interaction between sophisticated analysts and key decision makers. Although the CLIOSjre Process is designed to facilitate differences of opinion among different decision makers within an organization, in the public sector context decision making is likely to involve decision makers from the public implementing agency as well as from other major stakeholders with strong political, legal, financial, or other influence over the decision. This presents an institutional challenge that would need to be evaluated and considered carefully in the overall design of any new application of the CLIOSjre Process.

New Sources of Data. Perhaps the greatest strength of the CLIOSjre Process is its ability to systematically combine a variety of analyses into a single unified decision framework. As technology enables a more detailed and real-time understanding of the world, the CLIOSjre Process (and other multicriteria, multistakeholder decision aids) will help analysts integrate these new sources of data into existing decision-making processes in both the public and private sectors.

New sources of data and their associated analytical processes will likely provide new ways of evaluating the existing CLIOSjre metrics. Further research would be needed to resolve the appropriate methods for integrating these new sources of data into the CLIOSjre and other decision-making processes. Regardless of the particulars of integration, these new sources of data promise a more detailed understanding of our transportation investment decisions.

Acknowledgments

This research was supported by a 3-year research agreement with JR East. The authors are grateful to their collaborators at JR East for their input throughout the development process and their guidance on how to make the CLIOSjre Process useable for a transportation service provider. Thanks also to colleagues from the Regional Transportation Planning and High-Speed Rail Research Group at MIT for their critiques and suggestions throughout the development of this framework.

Author Contributions

The authors confirm contribution to the paper as follows: study conception and design: J.S., J.M., P.D.; data collection: J.M., P.D.; analysis and interpretation of results: S.M., P.D., J.M.; draft manuscript preparation: J.M. All authors reviewed the results and approved the final version of the manuscript.

References

1. Doyle, T. P. *Multicriteria Multistakeholder Decision Analysis: Applications to Transportation Planning.* Masters thesis. Department of Civil & Environmental Engineering, Massachusetts Institute of Technology, Cambridge, MA, 2016. http://web.mit.edu/hsr-group/documents/Thesis_Doyle.pdf
2. Sussman, J., R. S. Dodder, J.B. McConnell, A. Mostashari, S. Sgouridis, A. F. Archila, N. Jain, N. Stein, S. J. Carlson, M. Pea-Alcaraz, and R. Westrom. *The CLIOS Process: Special Edition for the East Japan Railway Company.* Massachusetts Institute of Technology, Cambridge, MA, 2014. http://web.mit.edu/hsr-group/documents/JRE_CLIOS_TeachingNote.pdf.
3. Sussman, J. *Transportation in the Northeast Corridor of the U.S.: A Multimodal and Intermodal Conceptual Framework.* Institution for Transportation Policy Studies, Tokyo, Japan, 2012. http://web.mit.edu/hsr-group/documents/jiti.pdf.
4. U.S. DOT Federal Railroad Administration. *NEC Future Tier 1 Final Environmental Impact Statement.* U.S. DOT Federal Railroad Administration, Washington, D.C. 2016. https://www.fra.dot.gov/necfuture/tier1_eis/feis/.
5. Bose, R. C. Mathematical Theory of the Symmetrical Factorial Design. *Sankhyā: The Indian Journal of Statistics (1933-1960)*, Vol. 8, No. 2, 1947, pp. 107–166.
6. Rolland, A. Reference-based Preferences Aggregation Procedures in Multi-criteria Decision Making. *European Journal of Operational Research*, Vol. 225, No. 3, 2013, pp. 479–486.
7. Mazurek, J. A New Model for Multiple Criteria Decision Making with Ordinal Rankings of Criteria and Alternatives. *International Journal of Economics and Statistics*, Vol. 2, 2014, pp. 75–82.
8. Porter, M. E. The Five Competitive Forces that Shape Strategy. *Harvard Business Review*, Vol. 86, No. 1, 2008, pp. 78–93.
9. Moody, J. *Development of a Predictive Coalition Building Analysis for Stakeholders of Sociotechnical Systems.* Masters thesis. Department of Civil & Environmental Engineering, Massachusetts Institute of Technology, Cambridge, MA, 2016. http://web.mit.edu/hsr-group/documents/Thesis_Moody.pdf.
10. de Neufville, R. Real Options: Dealing with Uncertainty in Systems Planning and Design. *Integrated Assessment*, Vol. 4, No. 1, 2003, pp. 26–34. http://dx.doi.org/10.1076/iaij.4.1.26.16461.
11. Andrews, K. R., C. R. Christensen, W. D. Guth, and E. P. Learned. *Business Policy: Text and Cases.* Irwin, Homewood, Ill., 1969.
12. Greco, S., M. Ehrgott, and J. R. Figueira. (eds.). *Multiple Criteria Decision Analysis: State of the Art Surveys.* Springer-Verlag, New York, 2016.
13. Susskind, L. *Good for You, Great for Me: Finding the Trading Zone and Winning at Win-Win Negotiation.* Public Affairs, New York, 2014.
14. Tufte, E. R. *Beautiful Evidence.* Graphics Press, New York, 2006.

The Standing Committee on Task Force on Data for Decisions and Performance Measures (A0030T) peer-reviewed this paper (18-00740).

TRR
JOURNAL OF THE TRANSPORTATION RESEARCH BOARD

Article

Transportation Research Record
2018, Vol. 2672(44) 93–102

DOI: 10.1177/0361198118788463
journals.sagepub.com/home/trr
SAGE

Applying Innovative Performance Metrics for Corridor Evaluation

Maria Vyas[1], Jeffery Harris[2], Ted Knowlton[3], G. J. LaBonty[4], Ronald T. Milam[5], Shawn Seager[6], Chris Williges[7], and Lisa Zundel[8]

Abstract

Widespread application of traditional performance metrics such as vehicle level of service has resulted in transportation networks that are often suited for only one mode of transportation to the detriment of others. Moreover, the continued investment in capacity expansion intended to meet level of service expectations increases financial burdens on transportation agencies for the capital, operating, and maintenance costs of these facilities, without regard to whether those resources are being used efficiently. Shifting focus to alternative and more comprehensive performance metrics can result in a more economical approach to transportation infrastructure, allowing communities to understand what investment combinations provide the best bang for the buck. Innovative performance metrics also allow communities to consider priorities beyond level of service, and design infrastructure accordingly. The Wasatch Front Central Corridor Study, undertaken in the Salt Lake City metropolitan area, is one example of a regional application of efficiency-focused performance metrics such as person throughput, seat utilization, benefit–cost ratio, access to employment, and economic impacts. Further research opportunities could include development of best practices to standardize benefit–cost analysis for transportation projects, other methods of evaluating overall system efficiency, and the role of parking policy and land use change in maximizing use of transportation resources.

Modern transportation planning has focused on the concept of reducing congestion and maintaining a free flow of traffic, and on measuring successes based on vehicle level of service (LOS). This paradigm began with the advent of the modern highway and interstate system, and was perpetuated as common practice through such manuals as the Highway Capacity Manual published by the Transportation Research Board (*1*). The result of this practice has been the continued expansion of roadways to attempt to maintain a stated LOS and the free flow and reliability desired by the traveling public. State departments of transportation are under ongoing pressure to manage these expectations while simultaneously outlaying large amounts of maintenance funds for the systems already built and requiring upkeep. Meanwhile, this approach has had deleterious effects on many other aspects of communities and transportation networks. Wider roads with faster traffic are significantly more dangerous for pedestrians (*2*). Car-centric networks sometimes completely lack any facilities for cyclists or pedestrians, forcing them to walk or ride on busy streets, on roadway shoulders or on landscaped areas or other facilities that are not accessible by those in wheelchairs or motorized scooters. Roadway expansion generally leads to higher traffic volumes, evidence of induced vehicle travel (*3*), which then leads to higher vehicle miles traveled (VMT) and the associated air pollutant and greenhouse gas emissions. Induced vehicle travel negates the congestion-mitigation benefits of the roadway expansion projects in a shorter time period than is generally expected. Research published by the Air Research Board of the California Environmental Protection Agency indicates that new lane capacity can be consumed by induced demand within a 10-year time period (with substantial consumption often occurring much sooner than that) although analysis forecasts for transportation projects are generally completed for a 30-year time horizon (*4*).

[1]Fehr & Peers, Salt Lake City, UT
[2]Utah Department of Transportation, Salt Lake City, UT
[3]Wasatch Front Regional Council, Salt Lake City, UT
[4]Utah Transit Authority, Salt Lake City, UT
[5]Fehr & Peers, Roseville, CA
[6]Mountainland Association of Governments, Orem, UT
[7]HDR, Inc., San Francisco, CA
[8]Utah Department of Transportation, Region Two, Salt Lake City, UT

Corresponding Author:
Address correspondence to Maria Vyas: m.vyas@fehrandpeers.com

As a result, U.S. communities experience a wealth of negative side effects in the service of a metric that may have outlived its usefulness in mature urban areas.

Shifting Focus to New Performance Metrics

Leaving LOS behind, many other performance metrics can rise up to take its place. Metrics such as person throughput and seat utilization allow transportation planners and decision makers to consider the overall efficiency of the transportation system, and remedy inefficiencies in that network through various management strategies. Application of a benefit–cost analysis (BCA) to transportation concepts can quantify the economic implications of various investments, helping planners and engineers understand the repercussions of construction actions. These and other metrics (such as VMT, collision rates measured over time, or level of traffic stress [*5*], among others) allow the transportation planning profession to move beyond LOS and consider community and infrastructure investments from a more comprehensive and economical perspective.

Case Study: The Wasatch Front Central Corridor Study

Origins of the Study

The area colloquially known as the Wasatch Front represents the entirety of the Salt Lake City, Utah metropolitan area, including over 60 individual municipalities from Ogden to Provo. This region has some of the highest growth rates in the nation: Utah is frequently among the top-growing states, ranked as the state with the highest growth rate between 2015 and 2016 (*6*). Statewide projections estimate that the population will increase from 3 million people in 2016 to 5.5 million people by 2065, a growth rate of 83%; it is anticipated that Utah's high growth rates will taper down from around 20.5% per decade in the 2015–2025 period to around 9% per decade between 2055 and 2065, representing a more moderate rate of growth (*7*). One million new people are anticipated in the urban areas of the Wasatch Front by 2040 (*8*), which is resulting in considerable pressure on decision makers to deal with various community needs such as transportation, education, and water resources.

Utah has a strong history of regional collaboration and futuristic thinking, and of public financial support for its transportation networks. The non-profit planning organization Envision Utah led efforts in the late 1990s to increase public awareness of regional planning issues and build support for infrastructure such as the regional light rail system, TRAX (*8*). The state legislature has passed multiple bills in recent decades authorizing significant infrastructure spending, and in 2015 approved a gas tax increase tied to fuel prices (*9*), to secure additional funding for transportation projects. The gas tax increase, introduced as House Bill 362 and signed into law in March 2015, places the additional revenues into the Transportation Fund, which is typically used to cover operational and maintenance costs for Utah Department of Transportation (UDOT) (*10*). Multiple rounds of voter referendums have resulted in local sales tax increases to support transit and trail infrastructure, although the 2015 general elections saw a failure to approve additional sales tax increases to support transit in the two most populated counties in the state (*11*).

Regardless of these successes, the region still struggles to pay for the maintenance of the infrastructure already constructed, much of which is not being used to its full capacity. Before 2008, UDOT pursued a "Good Roads Cost Less" philosophy, providing high levels of maintenance on its roads to prevent more costly repairs later; in 2008, the agency acknowledged that it did not have enough funding to continue this approach and instead moved towards a tiered prioritization system of funding highway operations and maintenance needs (*12*). In addition, a 2016 report from the Utah Office of the Legislative Auditor General warned that the Transportation Investment Funds typically earmarked for new capacity projects may need to shift to cover the growing costs of debt obligations and maintenance needs (*10*). Furthermore, the Utah Transit Authority (UTA), the public agency that builds, operates, and maintains light rail, commuter rail, and bus services for the urban area, was the subject of a legislative audit in 2014 that included a review of the agency's revenue and cost projections. The audit revealed that State of Good Repair costs for existing transit lines had not been effectively incorporated into long range planning documents, nor adequately accounted for in long range financial estimates (*13*). The agency has since taken steps to address the situation; for instance, in 2017 the agency planned to spend 32% of its annual budget on State of Good Repair projects throughout the system (*14*).

It was through this lens that the region's transportation agencies—UDOT, UTA, and the region's two metropolitan planning organizations (MPOs), Wasatch Front Regional Council (WFRC) and Mountainland Association of Governments (MAG)—decided to embark on the Wasatch Front Central Corridor Study (WFCCS). This effort was an acknowledgment by these agencies that continuing the current paradigm of heavy capital investments on Interstate 15 (I-15) between Provo and Ogden was neither efficient enough to meet growing travel demand nor financially or environmentally sustainable in the long term. The stated purpose of this study was to "develop a master plan of unique and innovative transportation strategies to address critical needs in the region's primary transportation corridor," to "evaluate all potential strategies, including transit, roadway, operations, policy, active transportation, freeway/surface street connectivity, roadway/transit connectivity, transit/ped/bike connectivity, etc." and to address "underutilized capacities, accommodation and management of future demand, system integration, and enhancements to maximize existing infrastructure" (*15*). The primary transportation corridor in question included the I-15 freeway, the TRAX light rail system, the FrontRunner commuter rail system, and several parallel arterials. The land use context for the primary corridor ranged from an urban central business district in downtown Salt Lake City to suburban communities outside the urban core, some of which have their own distinct downtown areas; the whole of the study corridor is fully developed.

To achieve all of these objectives, the study needed a new approach to performance measurement. Whereas LOS focuses on the automobile driver's perspective, this study needed to capture the perspectives of all transportation network users while also considering the outcomes the region desired (e.g., what did the region want to create, avoid, and protect). By measuring performance through the lens of efficiency, the definition of what constituted a transportation problem evolved. The LOS F conditions on I-15 could be expressed as a symptom of poor seat utilization caused by mispricing travel: freeway capacity is not managed in peak periods to incentivize carpooling or transit use, and individuals are inclined to make travel choices that are most convenient to them (i.e., driving alone) at the expense of the operations of the system. As a result, the study was able to focus on a wider range of solution strategies including demand management, behavior change, and capacity modification. Scenarios were developed to represent various locations on a transportation demand spectrum: one end of the spectrum would represent tight management of travel demand and a constrained network with very little new capacity added (if any), whereas the other end of the spectrum would represent adding significant new capacity for all modes. The three scenarios developed for WFCCS included a range of multi-modal solutions. Scenario 1 included refinement of the street grid, improved crossings over I-15 for all modes (but especially non-motorized modes), barrier-separated carpool and toll lanes on I-15, an extensive protected active transportation network, and free fares for transit. Scenario 2 further incentivized transit use beyond free fares to include doubling the frequency of the bus and light rail system, additional commuter rail improvements to increase speed and frequency, tolling of all freeway lanes in the peak period and direction, and a regional network of managed lanes to maximize directional capacity and prioritize transit, plus extensive transportation demand management programs. Scenario 3 focused on building capacity for all modes, with the addition of elevated freeway lanes, new commuter rail stations, increased capacity on arterials throughout the region, and construction of multiple mobility hubs at new or existing commuter rail stations. The metrics described in the following sections were calculated for all three scenarios for a 2050 horizon year, and were compared with both a 2050 business-as-usual scenario and a current-year scenario. Scenarios 1, 2, and 3 were ranked based on how well they performed on these and a range of other metrics, including travel time, travel time reliability, household travel costs, air quality impacts, VMT, and households within a walking or bicycling distance from transit. Many of these metrics provided insights on scenario performance but may be more familiar to readers than those described below. Although one scenario (Scenario 2) performed best on the most metrics, the agency partners preferred to recommend a hybrid of various scenario components rather than select only the components analyzed in Scenario 2. The hybrid scenario was integrated as a series of recommended projects into the next Regional Transportation Plan.

Application of Innovative Performance Metrics

The WFCCS developed and applied a wide range of transportation and economic performance metrics in the evaluation of the investment scenarios. These performance metrics fundamentally changed the definition of the problem and provided a new direction with regards to what would constitute a successful response to the problem. Although over 15 individual performance metrics were calculated for each scenario, several portrayed particularly meaningful differences between scenarios: person throughput, seat utilization, benefit–cost ratio, access to employment, and economic impacts such as personal income and employment. Evaluation of metrics like these, and the conversation among the transportation agencies around these metrics, were critical in moving the state of the transportation planning practice in Utah toward more comprehensive considerations of investment decisions. The WFCCS evaluation for the other performance metrics not described here (including freeway and transit peak travel time, travel buffer index, injuries and fatalities, VMT, peak and daily mode split, share of households within walking or cycling distance of high-quality transit, and transit access mode split) can be reviewed in the Wasatch Front Central Corridor Study Evaluation of Investment Scenarios Memorandum (*16*).

Person Throughput and Seat Utilization. Measuring person throughput and seat utilization addresses capacity and efficiency of both the freeway and the transit system, by comparing the number of people who move through the I-15 corridor at key screenline locations. These metrics demonstrate how well public infrastructure carries people (versus vehicles), and how efficiently transportation infrastructure is being used. For instance, observations for 2014 revealed that roughly 25–30% of all seats on the freeway in the WFCCS study area were occupied in the peak period and direction, leaving ample opportunities for improving the efficiency of the network. Transit seat utilization observations for the same time period revealed similar opportunities depending on the transit line under examination: some segments of the regional commuter rail line had more than half of their seats available in the peak direction and time period, although other light rail lines were filled beyond seating capacity. The transportation network also had capacity available in the off-peak direction and in the off-peak time periods.

The WFCCS methodology for evaluating person throughput and seat utilization for both current and future conditions involved using regional travel demand models, supported by Automated Passenger Counts (APC) data for the rail and bus transit networks and Performance Monitoring System (PeMS) data for the freeway network, and making separate calculations for each mode (*16*). To determine current person throughput on the transit network in the corridor, which includes both the TRAX light rail system and the FrontRunner regional commuter rail system, analysts evaluated APC boardings and alightings data at a station-by-station level to extract the total passenger load at screenline locations throughout the corridor.

Transit seating capacity was based on data provided by UTA for each transit mode, with assumptions made for the number of train car consists for peak period transit trips and the number of trips made during the peak period. For example, routes with a headway of 15 minutes were assumed to have 12 trips during the peak period. The number of trips was then multiplied by the seating capacity for each mode type provided by UTA. The transit person throughput was then divided by the total seats available to provide a current-conditions assessment of seat utilization percentage. Future-year scenario analysis of transit person throughput and seat utilization relied on regional travel model outputs of transit route ridership at the corresponding screenlines, applied to the transit project assumptions contained within the 2014–2050 Regional Transportation Plans for WFRC and MAG (*8*, *17*).

To determine freeway person throughput, the analysis assumed vehicle occupancy rates for freeway travelers based on observations and the 2012 Utah Household Travel Survey. PeMS data provided vehicle volumes at screenline locations throughout the corridor. Using the estimated occupancy rates discussed previously, freeway person throughput was calculated and then divided by the total seats to provide a seat utilization percentage. For the freeway analysis for future-year scenarios, model network links at these locations provided passenger car volumes and vehicle occupancies were taken from the travel demand model to calculate freeway person throughput and seat utilization.

Across the study area, the analysis of person throughput and seat utilization revealed trends occurring in each WFCCS scenario. For example, the future scenario with the highest amount of freeway construction (Scenario 3) moved the most people through the corridor, on the rail and freeway network combined. However, seat utilization in Scenario 3 was lower than in other WFCCS scenarios for both the freeway and the transit network because Scenario 3 did not have accompanying management strategies to maximize the efficiency of the system. Additionally, transit person throughput was lower under Scenario 3 than other scenarios, despite heavy investments in the rail network. The most management-focused scenario, Scenario 2, applied a combination of electrifying and double tracking the FrontRunner commuter rail system, free transit, increased transit frequencies, and freeway tolling. This resulted in the highest transit person throughput and highest freeway seat utilization (50% and upwards in some sections of the freeway). However, although Scenario 2 presented the highest transit person throughput, it did not have the highest transit seat utilization. In Scenario 2, the number of transit seats moving through the corridor doubled because of the increased transit frequency, so although more riders used the system, there were also more seats to fill. Compare this to Scenario 1, which applied some degree of freeway tolling and transit incentives to a lower-capacity transit network, resulting in more efficient use of the transit system.

The benefit to practitioners of evaluating corridor person throughput and seat utilization across modes is twofold. First, it considers how modes function together and either complement or detract from one another, which assists decision makers in considering how investing public funds in one mode's infrastructure system could have an impact on investments made for other modal systems. Second, it combines modes to evaluate overall corridor performance in terms of the number of people transported.

An issue raised in the analysis for WFCCS, which may be useful for future evaluation, is the role of induced vehicle travel in increasing person throughput. Scenario 3 (the high-investment scenario) had the highest amount of person throughput of the WFCCS scenarios, but it also had the highest modeled daily and peak period traffic volumes compared with other scenarios. The data suggests that the substantial freeway projects proposed in Scenario 3 had the effect of encouraging more traffic on the freeway, thereby raising the number of people moving through the corridor in automobiles. Scenario 3 also included capacity improvements such as grade separated interchanges on arterial routes parallel to I-15, which also saw a corresponding increase in vehicle traffic (for all scenarios, the regional arterial roadway network was represented in the modeling analyses, although reporting for metrics was focused on the I-15 corridor and parallel transit routes). Whereas more vehicle traffic has some undesirable effects such as increased VMT and associated vehicle emissions, the increase in capacity also suggests that residents of the region have increased access to less-expensive housing at the urban fringes. In comparison, Scenario 2 (the management-focused scenario) had lower traffic volumes and better travel times and reliability because of the management of the system; this suggests that the tolling features present in Scenario 2, combined with heavy transit investments and fare changes, had the effect of suppressing single-occupant vehicle trips, shifting trips onto transit, and increasing carpooling. The lesson learned in this example is to look carefully at metrics from multiple angles to better understand scenario performance, and avoid drawing conclusions from one metric alone. Readers should also know that a fixed person-trip table was used in the analysis of all scenarios; future practitioners may wish to take a different approach to person-trips.

Benefit–Cost Analysis (BCA). A BCA is a systematic approach to comparing the benefits and costs of investments throughout their lifecycles. The benefit-cost evaluations for WFCCS compare conditions in the "business-as-usual" scenario (Scenario 0) to WFCCS Scenarios 1, 2, and 3 (*18*). The BCA puts multiple measures of effectiveness into a single measure (benefit–cost ratio) to assess investment decisions. The benefit–cost ratio compares the benefits generated in each scenario to the capital and operating and maintenance (O&M) expenditures, and is a key measure for comparing the cost effectiveness of the WFCCS scenarios.

The BCA includes estimates for three types of costs over the analysis period: capital, roadway O&M, and transit O&M. The capital costs associated with each WFCCS scenario include the upfront design, engineering, and construction costs. Roadway O&M costs are associated with maintaining new roadways and bike paths, including pavement resurfacing and annual maintenance. Transit O&M costs are the incremental costs (such as employee wages, materials and supplies, and other operating expenses) of operating and maintaining new transit services.

Travel demand model outputs were used to estimate many of the benefits generated by each WFCCS scenario. These outputs included VMT by speed bin, vehicle-hours traveled (VHT) by speed bin, and the number of person-trips. These outputs were reported separately by vehicle type (i.e., auto or truck) and extrapolated over the entire analysis period. Transit benefits were estimated by mode using vehicle revenue-miles, vehicle revenue-hours, passenger-hours traveled (PHT), and the number of person-trips.

The outputs of the travel demand model were used to derive the economic benefits of each WFCCS scenario. Benefits were estimated separately for passenger and freight vehicles and various transit modes. Four categories of benefits were considered for the BCA:

- Travel time savings, representing the benefits of reduced travel times resulting from project improvements
- Vehicle operating cost savings, representing the benefits to roadway users from savings in fuel consumption and other out-of-pocket costs resulting from project improvements
- Crash cost savings, quantifying the number of crashes avoided as a result of project improvements
- Emissions cost savings, representing the benefits of reduced vehicle emissions in the study area

The BCA results over the analysis period are presented in Table 1. Presenting a monetized and indexed comparison of these benefits and costs for each scenario allowed decision makers to efficiently ascertain the order-of-magnitude difference between the scenarios. Moreover, with the understanding that a benefit-cost ratio of 1.0 represents a break-even point for a scenario, agency representatives could also quickly determine the degree to

Table 1. BCA Results

Project costs	Scenario 1	Scenario 2	Scenario 3
Lifecycle benefits ($ millions)	$3,194.2	$8,864.1	$1,156.6
Lifecycle costs ($ millions)	$910.2	$2,969.5	$3,326.5
Net present value ($ millions)	$2,284.0	$5,894.6	($2,169.9)
Benefit-cost ratio	3.51	2.99	0.35

which a scenario did (or did not) make financial sense from this perspective.

Access to Employment. This metric addressed the effect of each scenario on accessibility to jobs within a 45-minute travel time either by driving or by using transit. The ability of the transportation network to connect people to important destinations efficiently is one of its most important functions. Accessibility to more jobs provides benefits to both employers and employees. For employees, this means access to more employment opportunities as well as greater choice in the types of employment available. Access to larger employment opportunities, especially within a reasonable transit travel time, is particularly important for economically disadvantaged populations that may have no access or limited access to an automobile. A larger share of jobs within a reasonable transit trip time also makes transit more competitive, potentially lowering the use of single-occupancy vehicles, especially during peak times. For employers, the better the accessibility, the larger the labor pool. Demonstrating access to labor markets is key in maintaining and enhancing economic development in the region.

Regional travel demand models were used to estimate job accessibility within a 45-minute automobile or transit travel time for each scenario. Using the same travel time shed for both automobiles and transit was a deliberate choice: in one survey of Salt Lake City residents, 61% of respondents reported that they could take transit more frequently if they wanted to (*19*), suggesting that there is high potential to convince more travelers to use transit. Comparing transit to roadway accessibility at a parallel time shed (i.e., a 45-minute travel trip) allows analysts and decision makers to see the range of geography accessible by each, and test transit network improvements to see what level of change is necessary to make transit more competitive and attractive to new riders. Travel times for both freeway and transit were derived from the loaded model networks (i.e., networks where roadway and transit links have assigned volumes of passengers and vehicles). Accessibility was analyzed for economically disadvantaged or vulnerable populations, as well as the rest of the overall population. Agency staff provided information on "Communities of Concern," representing geographic areas that currently have higher shares of economically vulnerable populations. The travel demand models measured the weighted average number of jobs accessible in a 45-minute automobile or transit trip for traffic analysis zones (TAZs) that were considered part of a "Community of Concern," compared with all TAZs within the study area.

Table 2 provides the weighted average number of jobs accessible for vulnerable populations and the overall population within the study area TAZs for each scenario.

As shown in Table 2, Scenario 2 (the management-focused scenario) generally had the largest average number of jobs accessible for vulnerable members of the population as well as the population as a whole. For automobile trips this is likely to be the result of reduced congestion on I-15 caused by variable pricing in barrier-separated carpool lanes and pricing of general purpose lanes in the peak period and direction; for transit users this is an effect of faster commuter rail speeds and increased transit frequency throughout the network. Scenario 1 had the smallest effect on accessibility, because of its relative lack of roadway and transit capacity projects. Although Scenario 3 also improved job accessibility, the relative change was smaller, because travel times did not compare as favorably to Scenario 2 and a smaller geographic area was accessible within the defined time period as a result.

Readers will note that the average number of jobs accessible is higher for vulnerable members of the population than it is for the population at large; this is because the Communities of Concern that are home to these vulnerable populations are concentrated along the primary transportation corridor whereas more affluent communities are located some distance from the corridor.

Table 2. Average Number of Jobs Accessible for Households within the Study Area

	Scenarios	Accessible jobs (auto)	Accessible jobs (transit)
Vulnerable populations	Base year 2014	1,002,194	343,379
	Scenario 0	1,391,115	395,155
	Scenario 1	1,378,281	400,435
	Scenario 2	1,601,658	502,366
	Scenario 3	1,405,564	426,422
All populations within study area	Base year 2014	950,213	232,956
	Scenario 0	1,255,675	254,595
	Scenario 1	1,269,588	252,210
	Scenario 2	1,580,188	352,486
	Scenario 3	1,329,395	298,603

Therefore, residents in these Communities of Concern have faster access to the freeway and rail stations and can reach more jobs within the specific time window than some of their wealthier neighbors.

One benefit of applying an accessibility metric to specific populations is its ability to address social equity issues: how much value is experienced by the most in-need communities as a result of transportation investments, as compared with the rest of the population? The comparison of transit accessibility to automobile accessibility is also critical information, for two reasons. First, vulnerable populations are likely to be more dependent on the transit network for daily commutes and other transportation needs, and it is important to understand the range of geography and land use that can be accessed via the transit network. Second, to capture more riders on the transit system and achieve a WFCCS goal of improving mode balance, the transit system needs to be considered competitive with the roadway network if travelers are going to change their mode of commuting. Other research organizations and transportation partners have undertaken similar methods to evaluate accessibility issues, including the University of Minnesota Accessibility Observatory.

Economic Impacts. Evaluating economic impacts determined the changes in regional economic activity caused by the level of investment associated with each WFCCS scenario. These impacts occur as a result of changes in demand for goods and services resulting from construction expenditures and efficiencies gained from transportation improvements. They were measured using a REMI Policy Insight Model (which calculates employment by weighting full-time and part-time jobs equally). The total economic impacts of the WFCCS investment scenarios are the sums of three different economic effects:

- Direct effect: changes in economic activity as a direct consequence of the investments (e.g., transportation-related construction expenditures or savings in production costs because of transportation-related efficiencies)
- Indirect effect: changes in economic activity related to spending by industries that supply the ones experiencing the direct effects
- Induced effect: changes in economic activity related to employee spending (by employees of firms affected by the direct and indirect effects)

The WFCCS economic impact analysis considers two types of impacts:

- Short-term construction impacts related to the expenditure of transportation funding on construction and lifecycle O&M costs
- Long-term transportation efficiency impacts through production cost savings (meaning that vehicles circulating for business purposes can move more easily), amenity benefits (resulting from reduced emissions), and consumption re-allocation (related to fuel expenditures, so funds previously spent on fuel can be spent on other goods and services)

The economic impact analysis takes outputs from the BCA, and converts them into:

- Production cost savings (i.e., travel time savings, vehicle operating cost savings, and accident cost savings to trucks and business auto trips)
- Consumption re-allocation (i.e., fuel and non-fuel vehicle operating cost savings for non-business auto trips)
- Emissions amenity (i.e., emissions cost savings to trucks and automobiles)

These and other factors were used to simulate the economic impacts of the three WFCCS scenarios in the urban areas of the Wasatch Front (*18*). Economic impacts were assessed in terms of Gross Regional Product (GRP) (i.e., total value added within the region including employee compensation, proprietor's income, and other property type income), personal income (i.e., total earnings from work, investments, and other sources), and employment (i.e., total number of full-time and part-time jobs). Table 3 presents the results of the economic impact analysis.

The results show that the magnitude of impacts is similar under Scenarios 2 and 3, but Scenario 1 yields significantly lower impacts (less than half of Scenario 2 or Scenario 3) because it has considerably fewer projects for construction during the study period. Over the 2025–2050 period, the cumulative impact on GRP is estimated at $5.8 billion for Scenario 1, $14.7 billion for Scenario 2, and $11.8 billion for Scenario 3 (see Table 3). The employment impact is largest in Scenario 3, even though the impact on GRP is nearly 20% lower than Scenario 2. This is because nearly all the impacts in Scenario 3 are created by construction, which is very labor intensive. As a result, the output per employee is relatively low, because construction work requires more employees to produce the same unit of output as other industries. It should additionally be noted that, whereas construction jobs are counted as an economic benefit, members of the traveling public may not necessarily perceive construction as a benefit because of its temporary impact on their daily commutes.

Addressing economic impacts with this approach was valuable because it allowed the comparison between

Table 3. Economic Impact Analysis Results

Impact metric	Scenario 1	Scenario 2	Scenario 3
Total gross regional product ($ millions)	$5,833	$14,735	$11,831
Construction impact	$3,012	$9,320	$10,984
Transportation efficiencies	$2,821	$5,416	$847
Total personal income ($ millions)	$4,652	$10,650	$10,244
Construction impact	$2,874	$7,580	$9,605
Transportation efficiencies	$1,779	$3,071	$639
Total employment (job-years)	44,242	103,263	103,615
Construction impact	28,236	77,148	96,974
Transportation efficiencies	16,005	26,115	6,640

Note: A job-year is simply defined as one (part- or full-time) job for a year. The sum of the construction and transportation efficiencies metrics may not equal the total because of rounding.

economic gains associated with construction jobs and economic gains associated with transportation efficiency. This comparison facilitated discussion and enhanced understanding among decision makers about the relative value and longevity of those gains, raising questions of whether the short-term jobs gain of large infrastructure projects was an appropriate argument for such construction. Decision makers noted that, to maintain the gains produced by construction impacts, new similar-scale infrastructure projects would need to come online as current projects neared completion. If new projects were not to replace the old ones, construction jobs generated as a result of the scenario would disappear on completion of the scenario improvements.

Lessons for Practitioners and Future Research Opportunities

Lessons Learned

Several lessons were learned through this process that could be valuable for practitioners as well as stakeholders and decision makers responsible for funding capital transportation investments, especially with additional research applied. One lesson that applies to all the metrics used is that all forecasts and assumptions carry a degree of risk and uncertainty. Every WFCCS scenario that was analyzed contained a set of assumptions about how various components would operate (toll rates, transit fares, or access locations to barrier-separated freeway lanes, for example), which were coded into a regional travel model that contains its own set of assumptions about human travel behavior. These assumptions affect the performance of the scenarios on various metrics, and could potentially have biases encoded into them as well. The analysis also assumed that the America of tomorrow will largely resemble the America of today, which is unlikely to be true. Future studies of large-scale regional transportation plans should incorporate sensitivity analysis to account for a range of possible futures, allowing decision makers to see how resilient our transportation investments might be in the face of various future outcomes.

Selection of appropriate performance metrics for the region, agencies, and corridor in question requires a thoughtful and methodical approach. Whereas analysts may enjoy the exercise of comparing abundant sets of performance metrics, decision makers (often without extensive technical training or knowledge and with limited time to address planning issues) will be required to sort out the meaning behind the various metrics. Practitioners should resist the temptation to explore a wide range of metrics and should instead focus on metrics that are the most meaningful to stakeholders, while demonstrating differences between alternatives. In the case of WFCCS, over 15 individual metrics were analyzed, yet only a handful (several of which are discussed in this paper) spurred meaningful dialogue about the choices presented. Many metrics showed no discernable difference between WFCCS scenarios, particularly those analyzing conditions in which travel demand model performance was inadequate (for example, changes in active transportation mode split between scenarios). In this instance, it was only possible to know that there was no discernable difference in some metrics because those metrics had all been calculated and compared; future practitioners may wish to pre-screen some metrics in other evaluations before a full calculation to gauge whether the effort to calculate results is likely to be worthwhile, or whether time and energy is better spent on evaluating other metrics.

Another challenge with using performance metrics in WFCCS was a perception of bias. With a sponsoring group composed of the DOT, the transit agency, and two MPOs representing areas with different political and philosophical tendencies, it was difficult to identify metrics for evaluating scenarios that would be acceptable to all parties involved. Partner agency staff agreed that goals and metrics should be mode-neutral: for instance, a stated goal of "Increase transit use" has a bias towards a particular mode, but a stated goal of "Increase person throughput" allowed all modes to be considered together as an integrated and complementary system in the corridor, rather than pitting transit against highway to see which fares better. Some performance metrics are very challenging to make mode-neutral. For example, a goal and associated performance metric related to reducing VMT is going to favor transit-friendly solutions over single-occupancy vehicle solutions; therefore it is important to recognize that some critical performance metrics cannot eliminate inherent bias or a limited perspective. However, having the conversations about how to remove mode bias as much as possible can go a long way toward achieving agency buy-in on selected goals and performance metrics.

Selecting which performance metrics to evaluate also tends to indirectly define the transportation problem to be solved and the range of potential solutions that can be developed for a corridor. For instance, goals and performance metrics related to improving air quality are going to favor solutions like increased transit capacity, active transportation networks, and carpool incentives. Similarly, performance metrics like reduced vehicle travel time would generally result in roadway widening, intersection capacity increases, and access closures. Agencies should carefully consider the nature of the problem to be solved and select performance metrics accordingly, rather than allowing the selection of metrics to dictate the issue and the range of solutions to the problem.

Future Research Opportunities

Several questions raised by WFCCS could be useful research topics for future evaluation. These are summarized briefly below:

- Clarify and standardize the most appropriate set of calculations to use in a BCA, to reduce confusion about what is being analyzed and compared.
- Explore the potential role of land use change in affecting travel demand, especially the degree to which future land use changes might accommodate travel demand if no more transportation infrastructure investments were possible.
- Analyze the potential of widespread parking policy changes in the urban core to affect travel behaviors at a regional scale; in locations similar to the Wasatch Front, where downtown parking is ample and often free, a change of parking policies by the capital city could have a dramatic effect.
- Study the roles of connected and autonomous vehicles (CAVs) in the transportation network, with a focus on the likely fleet mix at various time intervals, to better plan for needed investments in infrastructure.

Author Contributions

The authors confirm contribution to the paper as follows: study conception and design: Jeff Harris, Ted Knowlton, G.J. LaBonty, Ronald Milam, Shawn Seager, Lisa Zundel, Maria Vyas; data collection: Chris Williges, Maria Vyas; analysis and interpretation of results: Maria Vyas, Ronald Milam, Chris Williges; draft manuscript preparation: Maria Vyas, Chris Williges, Ronald Milam, G.J. LaBonty. All authors reviewed the results and approved the final version of the manuscript.

References

1. *Highway Capacity Manual: A Guide for Multimodal Mobility Analysis*, 6th ed. Transportation Research Board, Washington, D.C., 2016.
2. Bartolomeos, K., P. Croft, S. Job, M. Khayesi, O. Kobusingye, M. Peden, D. Schwebel, D. Sleet, G. Tiwari, B. Turner, G. van Waeg. *Pedestrian Safety: A Road Safety Manual for Decision-Makers and Practitioners*. World Health Organization, Geneva, 2013, pp. 31–32.
3. Duranton, G., and M. Turner. The Fundamental Law of Road Congestion: Evidence from US Cities. *American Economic Review*, Vol. 101, No. 6, 2011, pp. 2616–2652.
4. Handy, S., and M. Boarnet. *Impact of Highway Capacity and Induced Travel on Passenger Vehicle Use and Greenhouse Gas Emissions*. Policy Brief, California Environmental Protection Agency Air Resource Board, 2014, pp. 3–4.
5. Mitchell, C., and R. Milam. *PAS Memo: Smart Transportation Metrics for Smart Growth*. American Planning Association, Chicago, IL, 2016, pp. 2–6.
6. United States Census Bureau. *Utah is Nation's Fastest Growing-State, Census Bureau Reports*. Release Number CB16-214. United States Census Bureau, 2016.
7. Hollingshaus, M., E. Harris, C. Jeppsen, and P. Perlich. *The Beehive Shape: Provisional 50-Year Demographic and Economic Projections for the State of Utah, 2015–2065*. Policy Brief, University of Utah Kem C. Gardner Policy Institute, 2016.
8. Wasatch Front Regional Council. *Regional Transportation Plan 2015–2040*. Wasatch Front Regional Council, Salt Lake City, Utah, p. 14.
9. Davidson, L. Utah Legislature: Compromise increase in gas tax OK'd at eleventh hour. *Salt Lake Tribune*, March 13, 2015.
10. Office of the Legislative Auditor General. *An In-Depth Budget Review of the Department of Transportation*. Report to the Utah Legislature Number 2016-05. State of Utah, 2016, pp. 3–4, ii.
11. Davidson, L. Prop 1 Dies in Salt Lake County; Margin of Defeat Widens after Canvass. *Salt Lake Tribune*, November 17, 2015.

12. Utah Department of Transportation. *Strategic Direction*. Utah Department of Transportation, 2016, p. 13.
13. Office of the Legislative Auditor General. *A Performance Audit of the Utah Transit Authority*. Report to the Utah Legislature Number 2014-06. State of Utah, 2014, pp. 47–61.
14. Utah Transit Authority. *2017 Operating Budget, Exhibit A*. Utah Transit Authority, 2016.
15. Utah Department of Transportation. *Standard Request for Qualifications: Project No. F-R299(198); I-15/FrontRunner Corridor Transportation Plan*. Utah Department of Transportation, 2014.
16. Fehr & Peers. *Wasatch Front Central Corridor Study Evaluation of Investment Scenarios Technical Memorandum*. Utah Department of Transportation, Salt Lake City, UT, 2017, pp. 13–16.
17. Mountainland Association of Governments. *TransPlan40 Regional Transportation Plan: 2015–2040 Plan for the Provo/Orem Metropolitan Area*. Mountainland Association of Governments, Orem, UT, pp. 42–43.
18. HDR. *Wasatch Front Central Corridor Study Economic Impact Analysis Technical Memorandum*. Utah Department of Transportation, Salt Lake City, UT, 2017, pp. 13–22.
19. Outwater, M., B. Sana, N. Ferdous, J. Lobb, and B. Woodford. *TCRP Report 166: Characteristics of Premium Transit Services that Affect Choice of Mode*. Transportation Research Board of the National Academies, Washington D.C., 2014.

The Standing Committee on Performance Management (ABC30) peer-reviewed this paper (18-03324).

Article

TRR
JOURNAL OF THE TRANSPORTATION RESEARCH BOARD

Transportation Research Record
2018, Vol. 2672(44) 103–113

DOI: 10.1177/0361198118791910
journals.sagepub.com/home/trr
SAGE

Long-Distance, Multimodal Freight in a Continental Transportation Model

Klaus Noekel[1], Ian Williams[2], and Davide Fiorello[3]

Abstract
This paper proposes a model of long-distance freight traffic that is suitable for transportation models covering very large areas. Three challenges are discussed in turn. First, the geographic distribution of trips not only depends on locations of production and consumption, but also on the choice between alternative logistic distribution chains and the locations of intermediate distribution centers. Second, any stage of the distribution chain may combine several modes into a multi-leg transport. Third, the large scale of the model leads to large zones, implying a significant share of intrazonal traffic. The proposed approach adapts the four-stage model by generalizing destination choice into a distribution channel model and by introducing a mode sequence choice model for multimodal transport. A simplified distance band model is applied to intrazonal traffic.

TRIMODE is a strategic transport network model simulating passenger and freight transport activity and covering all mobility—urban, regional, national, international and intercontinental—for the European Union and neighboring countries, fully integrated with an economy and an energy model. The tool is being developed to provide the European Commission with a usable and reliable tool to support the assessment of transport policies and infrastructure investments.

Even though its focus is on mobility at the European national and international level, TRIMODE is targeted at simulating the whole transport demand, from local to intercontinental, for both passenger and freight. To deliver good quality forecasts and responses to policy simulation, TRIMODE aims to provide a realistic representation of how transport activity is generated and occurs on European territory. Freight transport activity is rooted in interregional trade but its pattern is heavily influenced by logistic operations. Economic transactions between zones give rise to movements of goods, which are transported through various distribution chains, from the original producer to the final consumer. In many cases, different transport modes are used to move goods within a specific distribution chain, so that multimodality is a common feature of freight transport. In many distribution chains the final leg consists of delivering relatively small amounts of products to retailers. A large share of these movements occurs at distances that, given the spatial granularity of a model like TRIMODE, are internal to each zone (TRIMODE works at NUTS 3 level. NUTS [nomenclature of territorial units for statistics] is a hierarchical system for dividing up the economic territory of the EU. NUTS 3 level corresponds to the finest level available: there are 1,342 NUTS 3 zones in the EU. Details can be found in *NUTS: Nomenclature of Territorial Units for Statistics* [*1*]).

Accordingly, many extra challenges arise when developing operational, comprehensive models that represent all freight movements on all modes across a large number of countries. To address these in TRIMODE, distribution chains are explicitly represented in the modeling of freight demand patterns, whereas multimodal assignment is used to estimate freight movements within each chain. Furthermore, local distribution movements occurring within zones, even though they are not assigned to the network, are modeled in detail by means of a distance band approach.

This paper presents how these three aspects are implemented in TRIMODE, starting in the next section with the modeling of freight distribution chains, then the description of multimodal assignment and finally, the treatment of intrazonal movements. A final section presents some conclusions.

[1]PTV Group, Karlsruhe, Germany
[2]TRT Trasporti e Territorio, Cambridge, UK
[3]TRT Trasporti e Territorio, Milano, Italy

Corresponding Author:
Address correspondence to Klaus Noekel: klaus.noekel@ptvgroup.com

Freight Distribution Channel Modeling

When examining wide-ranging transport policy measures at the European scale it is important to ensure that freight is represented as a derived demand. The pattern of goods movement arises from the interplay between:

- The spatial pattern of the supply (production) and demand (consumption) and of the cost of goods
- The logistics structures that are in place to ship goods from producers (supply) to consumers (demand), via distribution centers (DCs), using the most suitable combination of modes

The *economic transactions* between suppliers and consumers, and the *logistics operations* that actually deliver the physical goods, are the two main drivers behind the observed pattern of freight movements. Accordingly, we draw a distinction between two separate matrices of movements that are used conceptually in demand analysis for freight transport:

- The pattern of economic trade in goods from their initial producer to their ultimate consumer; this is called the production–consumption (P-C) matrix of trade. Changes in this matrix are strongly influenced by economic changes *outside* the transport and distribution sectors. This matrix is forecast within the TRIMODE Economy model, a full-fledged macroeconomic and regional model based on computable general equilibrium methodology. Its structure is based on the GEM-E3 model, documented in (*2*). This component is not further discussed in this paper.
- The actual set of physical transport movements generated by the logistics structure that distributes and transports these P-C trades in practice; this is called the origin–destination (O-D) matrix of shipments and is forecast within the TRIMODE Freight Demand model using the distribution channel choice methodology described below. Changes in this matrix are strongly influenced by changes *within* the freight transport and the distribution industry sectors.

Representing Logistics and Distribution Channels

The formulation of the distribution channel choice mechanism within the freight demand model uses two distinct but intertwined structures that are solved simultaneously within its mathematical formulation:

- A *spatial structure* generating transport movements for traded goods from production zones to consumption zones. This is a conventional component of standard transport models. The corresponding monetary flow to pay for the goods transported passes in the reverse direction: from the consumption zone to the production zone.
- A *distribution channel* connecting a specific type of good from its original zone of production through DCs to its ultimate zone of consumption, being transported via various types of intermediate logistic legs.

The distribution channel representation that is used was originally developed in 1978 for a freight model for the state of Sao Paulo in Brazil (*3*), then refined in UK government funded research studies (*4*), and was subsequently successfully implemented at a large scale in 2010 in the Base Year Freight Matrix (BYFM) model for all domestic freight transport movements on road and rail within Great Britain (*5*). The distribution channel choice mechanism in TRIMODE refines and simplifies that used previously in BYFM.

A 2015 review (*6*) of the consideration of logistics within over 100 freight models found that few operational, large-scale models represent "distribution logistic hubs" (i.e., distribution channel choices), whereas even fewer represented "transport logistic hubs" (i.e., multimodal assignment). Accordingly, TRIMODE is unusual in integrating a detailed, though aggregate, representation of both of these features across a wide range both of logistics options and of combinations of freight modes.

The SAMGODS Swedish national freight model and its sister model for Norway (*7*) represent both distribution channel and multimodal choice using a disaggregate modeling approach. In the United States, the Freight Activity Microsimulation Estimator model developed by the University of Illinois also represents both features (*8*). Because of the client's requirement that TRIMODE be developed across a large number of countries, while using only data already available (invariably aggregate data), neither of these alternative approaches would have been feasible to implement on such data.

We start by introducing some terminology and context:

- The ultimate consumption zone c in this distribution channel context denotes the zone (NUTS 3 region), at the end of a particular distribution channel, in which either the *intermediate or final consumption* of that good takes place.
- The original production zone p is the zone (NUTS 3 region) at the start of a particular distribution channel. It is not one of the DC zones at which goods may be transferred or stored along the distribution channel prior to reaching their ultimate consumption zone.

The equations that estimate the transport flow T^r_{pc} from the production zone p to the consumption zone c for type of freight r, take explicit account of intermediate distribution activities. Figure 1 illustrates this conceptual shift from the left-hand box (the P-C presentation) to the middle box, which represents the actual freight movements on a logistic leg-by-leg basis, through the stages of the distribution chain. The transport flow T^r_{pc} is replaced by a more general procedure that adds a further layer to the model formulation. This extra layer builds up the transport flow in a manner that reflects the logistics options that are available at each distribution stage, connecting the original production zone through to the ultimate consumption zone for a specific commodity, through use of a succession of intermediate logistic legs along the distribution chain, as illustrated in Figure 1. The right-hand box is the equivalent channel-based representation of the leg-based middle box. It illustrates the three separate distribution channels that a consignment could take through this example distribution chain.

The logic behind this approach is now illustrated in the context of the example in Figure 1:

- The ultimate demand for consumption of a good of type r by households in zone c is satisfied by receiving these goods via some local retailer, denoted by logistic leg (A).
- These goods arrive at the retailer through a number of different types of logistic legs, either:
  - *directly from a producing factory*—denoted by logistic leg type (D), or
  - *indirectly from a distribution center* in which the goods have been warehoused or cross-docked between logistic legs—denoted by logistic leg types (B) and (C).
- For those goods that are output from the DC B, they will first have needed to be delivered there from DC A, using logistic leg type (E).
- The solution algorithm continues moving up the distribution chain through the sets of distribution centers for all zones until all distribution chain legs (in this particular example just the further logistic leg type [F]) have been traversed back to the original production zones (the blue box denoting the depot of the factory at which the goods were first produced).

This approach is generalized into a procedure for a given type of good that commences with the set of ultimate consumption zones at the top of the right-hand box in Figure 1 and then successively satisfies their demands by moving down each of the distribution channels, transporting goods from producers or DCs, via the appropriate logistic leg type and intermediate zone. The procedure continues until all demand for goods has been

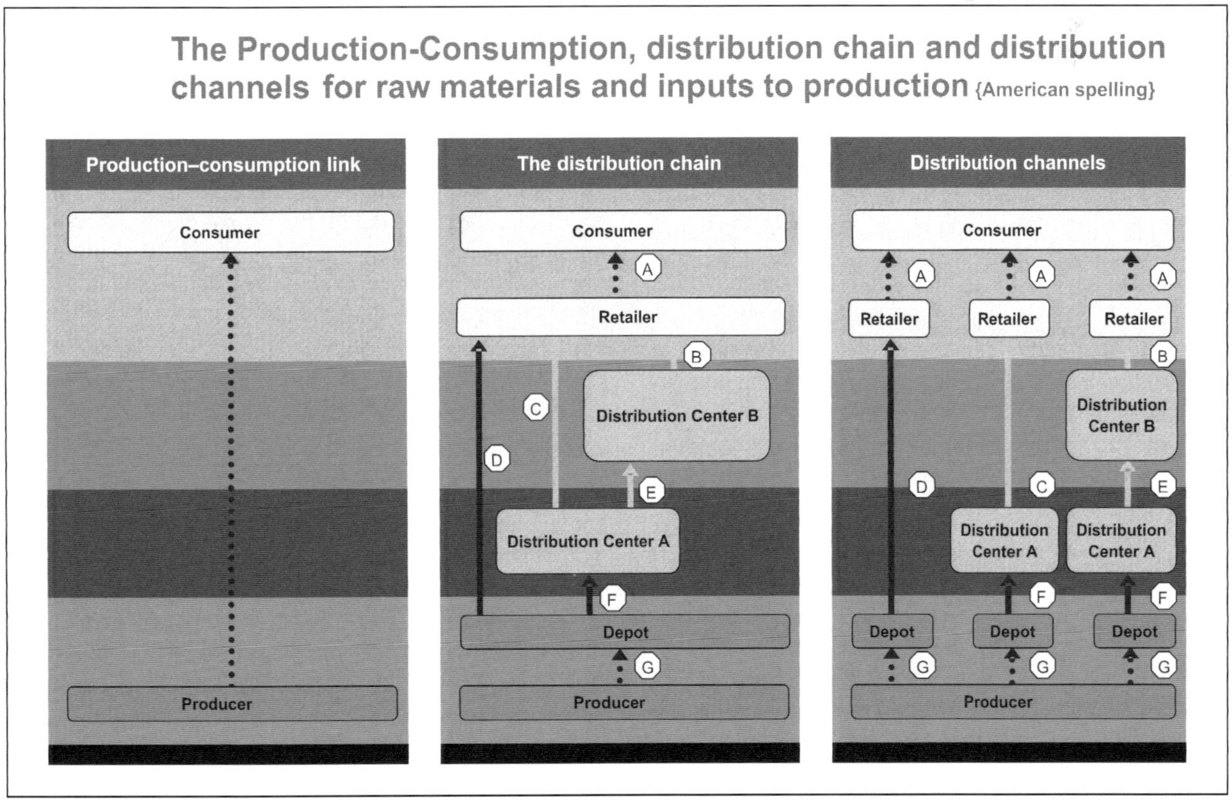

Figure 1. The production–consumption link, distribution chain, and distribution channels.

satisfied by the output from production or from import zones at the bottom of the box.

In practice, the computation is more complex because for each type of distribution chain in Figure 1 there will be many DCs and producers competing in different zones, so that the range of distinct options increases greatly. However, the dimensional scale of this solution algorithm is not overwhelming, provided that within a DC the goods are pooled and that the exact consumption zones of each output shipment further up the distribution chain do not need to be remembered within the solution algorithm when selecting DCs or production zones lower down in the chain.

Formulation of the Distribution Channel Choice Model

The mathematical formulation below specifies the integration of the spatial distribution channel choice and its associated logistic leg choice within the TRIMODE Freight Demand model component.

The solution algorithm adopted is the distribution channel choice approach shown in the right-hand box in Figure 1, rather than the distribution chain (leg choice) approach. It commences at the consumer end and traces back through each distribution activity along the distribution channel to end at the producer. The channel stages and logistic leg types at the top and bottom of each distribution channel are predominantly local short-distance trips that will be represented as intrazonal movements (see separate section below).

The following is the starting point for the calculations of the amount moved on each logistic leg.

D_c^r denotes the total ultimate amount of consumption of freight type r within the consumption zone c. It is based on forecasts from the TRIMODE Economy model.

X_{jk}^{rh} denotes the amount of the freight type r delivered to zone j via the distribution channel of type h in the kth stage back from the ultimate destination. It is calculated at stage $k = 1$ as:

$$X_{j1}^{rh} = p_{h|rj} D_c^r \quad \forall j = C, r \tag{1}$$

In this ultimate consumption zone only, implement a discrete choice model that is to be calibrated to match the handling factor value (handling factor denotes the average number of logistic legs for that freight type) that is exogenously specified for each specific freight type. This probabilistic choice is between competing distribution channel types h and is based on their relative costs of moving goods through to that consumption zone from all of the various available original production zones:

$$p_{h|rj} = e^{-\lambda^r\left(\tilde{u}_{j1}^{rh} + r_j^{rh}\right)} \Big/ \sum_{h'} e^{-\lambda^r\left(\tilde{u}_{j1}^{rh'} + r_j^{rh'}\right)} \quad \forall j = C, r \tag{2}$$

Each of these distribution channel types h has its own exogenously specified:

- Total number of logistic stages, each representing one of a succession of intermediate DCs or other facilities that need to be traversed
- Sequence of the type of storage facility used in each successive stage k, starting with $k = 1$ from the ultimate consumption zone end and increasing the stage count when traversing the distribution channel in the direction back to the original production zone
- Logistic leg type l (e.g., primary, secondary, or tertiary, etc.) associated with each movement stage k, between each of its successive pairs of facilities

r_C^{rh} denotes the unit residual disutility, if any, within the ultimate consumption zone c. If there is insufficient data to calibrate these values zonally using observed totals as constraints to be matched, the zonal subscript is dropped and the term acts as an exogenously input alternative specific constant for each distribution channel.

λ^r is a calibrated parameter that denotes the extent to which deliveries of freight type r concentrate on those distribution chains that can deliver goods with the lowest generalized cost of sourcing.

$\tilde{u}_{j1}^{rh}$ (see Equation 12).

The next component of the model is the spatial distribution of the leg l among the competing origin zones.

D_{jk}^{rhld} denotes the amount of freight type r delivered to the distribution activity d in zone j, via the logistic leg of type l in the kth stage back from the ultimate destination on the distribution channel of type h:

$$D_{jk}^{rhl(d)} = a_k^{rhld} X_{jk}^{rh} \quad \forall d, j, h, k, l, r \tag{3}$$

a_k^{rhld} is an (I-O) coefficient which takes a value 1, if the logistic leg type l is used on the stage k of the distribution channel h, otherwise it takes the value 0. It denotes the type of incoming logistic leg type l arriving at distribution activity type d, that was used to deliver this freight type r from its previous production or distribution stage.

For any given type of distribution activity d, only a subset of all logistic leg types l, can deliver to it.

X_{ik}^{rh} denotes the total volume of the freight type r that is transported out from the DC or factory in zone i via the $k-1^{st}$ stage of distribution channel h, to meet all of the demands for delivery elsewhere. It is calculated for $k > 1$ as:

$$X_{ik}^{rh} = \sum_{jld} T_{ijk-1}^{rhl} \quad \forall i, h, k, r \tag{4}$$

T_{ijk}^{rhl} denotes the volume of the freight type r that is delivered on a logistic leg of type l in stage k of distribution channel h, from a DC or production factory in the zone i to meet the demand for consumption within the distribution activity d in zone j.

In principle, this is an O-D matrix, rather than a P-C matrix, though those movements that go directly from a producer activity (i.e., zone $i = p$) to an ultimate consumer activity (zone $j = c$) will be common to both the O-D and P-C matrices:

$$T_{ijk}^{rhl} = D_{jk}^{rhl(d)} p_{i|rjhkl} \quad \forall i, j, h, k, l, r \tag{5}$$

$p_{i|rjhkl}$ denotes the probability that a consignment of the freight type r that arrives at zone j will have been delivered there from zone i. This origin zone choice probability is calculated using a standard logit type discrete choice model as:

$$p_{i|rjhkl} = S_i^{rl} e^{-\lambda^{rl} u_{ijk}^{rhl}} / \sum_{i'} S_{i'}^{rl} e^{-\lambda^{rl} u_{i'jk}^{rhl}} \quad \forall i, j, h, k, l, r \tag{6}$$

S_i^{rl} is a zonal size term that denotes the capacity of the origin zone i for the delivery of the freight type r. Depending on the type of logistic leg l, it will represent either the size of the relevant warehousing in the DCs in the zone, or the production capacity of the factories producing the good there.

u_{ijk}^{rhl} is the cumulative disutility (Equation 9) of delivering a unit of freight type r to zone j on logistic leg type l in stage k of distribution channel h that has been sourced from the origin zone i.

λ^{rl} is a calibrated parameter that denotes the extent to which deliveries of freight type r on logistic leg type l concentrate from those origin zones that can deliver goods with the lowest generalized cost of sourcing. It is calibrated to match the observed pattern of lengths of haul.

This market clearing equation completes the representation of distribution channels:

$$\sum_p X_p^n = \sum_{ce} D_c^{ne} \quad \forall n \tag{7}$$

The aggregate totals of the P-C matrix are calculated by accumulating the successive O-D delivery legs across the set of distribution channels h, the different combinations of intermediate logistic legs of type l that may be used to travel between a variety of intermediate pairs of zones:

$$T_{pc}^{r*} = T_{pc1}^{ru4} + \sum_j \left(T_{jc1}^{rv3} p_{p|rjv21} \right) + \sum_j \left(T_{jc1}^{rw2} \sum_i p_{i|rjw22} p_{p|riw31} \right) \forall c, p, r \tag{8}$$

Here the terms on the right-hand side of Equation 8 denote that there are three competing distribution channels as in Figure 1 above. It can be extended to represent longer distribution channels.

Transport disutilities (generalized costs) within distribution channels, are specified as:

u_{ijk}^{rhl} denotes the cumulative *zone pair consumption disutility* (generalized cost) of sourcing from the origin zone i one unit of the freight type r for delivery to zone j, within stage k of the distribution channel h. This comprises: the disutility of the previous delivery to origin zone i, plus the cost of production or of storage at the DC within the zone i, plus the subsequent disutility of transporting the goods using a logistic leg of type l through to the destination zone j. It is represented as:

$$u_{ijk}^{rhl} = w_{ik+1}^{rh} + \hat{u}_{ij}^{rl} \quad \forall i, j, h, k, l, r \tag{9}$$

$\hat{u}_{ij}^{rl}$ denotes the *transport disutility* associated with transporting one unit of the freight type r from a DC or a factory in zone i to delivery zone j on a logistic leg of type l. This transport disutility is a function of the monetary cost, time, quality of service, and residual disutility for the modes between the pair of zones. It is output from a standard hierarchical logit mode choice model as a composite logsum across the set of competing main modes for the zone pair along their equilibrium paths through their multimodal network, computed as described in the next section. For intrazonal legs, transport disutility depends on the distribution of demand among different distance bands, as explained in the penultimate section below.

The zonal *distribution disutility* (at the factory or origin warehouse gate) of one unit of the freight type r that is available in the zone i for delivery using stage k of the distribution channel of type h is:

$$\begin{aligned} w_{ik}^{rh} &= \hat{c}_i^{rd'} + \hat{r}_i^{rd'} \quad \forall i = p, h, l(d'), k, r \quad \text{for initial production zones} \\ w_{ik}^{rh} &= \hat{c}_i^{rd'} + \hat{r}_i^{rd'} + a_k^{rhld'} \tilde{u}_{ik}^{rh} \quad \forall i, h, l(d'), k, r \quad \text{for other DC zones} \end{aligned} \tag{10}$$

Here the notation $l(d)$ denotes the particular distribution activity of type d' at which the logistic leg of type l terminates.

$\hat{c}_i^{rd'}$ denotes the input factory gate price of the good if it is the starting logistic leg, that is, from the original production zone, $i = p$. Otherwise it denotes the input unit cost of warehousing the good for the distribution activity of type d' within dispatch zone i.

$\hat{r}_i^{rd'}$ denotes the unit residual disutility, if any, of the good, either at the factory gate if it is in the starting logistic leg, or otherwise for warehousing the good.

In all base year and forecasting scenario runs, the total production in each zone is applied as an aggregate production constraint that needs to be matched within the converged solution. This constraint is matched through use of Equation 11 to calculate the required disutility value that is then passed up through the distribution channels so as to adjust the allocation of supply by zone. The notation "$hk \in p$" indicates summation over all of the distribution stages that emanate from that original production zone p for the freight type r:

$$\hat{r}_{ps}^{rd'} = \hat{r}_{ps-1}^{rd'} + \left(\frac{1}{\lambda^r}\right) \ln\left(\sum_{ehk \in p} \frac{X_{pks}^{reh}}{X_p^r}\right) \quad \forall p, r,$$
$$\text{d'} = \text{producer/factory} \qquad (11)$$
$$\hat{r}_{p0}^{rd} = 0$$

The model is solved iteratively, as indicated by the iteration subscript s that has been explicitly included in the relevant terms in Equation 11. The endogenously calculated variables in all of the other equations above should equally include this model iteration subscript s, however, to avoid representation complexity it has not been explicitly presented within them.

An analogous residual disutility procedure is used to ensure that the input P-C tonnages by freight type and zone pair are also matched.

Each of the logistic legs of type l that is used to replenish the stocks of the good in a DC in zone i, needs to be sourced from a set of origin zones q, each with the corresponding cumulative zone pair consumption disutility: u_{qik}^{rhl}. The expected minimum disutility over all possible origin zones q, is given by the following logsum calculation:

$$\tilde{u}_{ik}^{rh} = \left(-\frac{1}{\lambda^{rl}}\right) \ln\left(\sum_{q} S_q^{rl} e^{-\lambda^{rl} u_{qik}^{rhl}}\right) \quad \forall i, h, k, l \in (h, k), r \quad (12)$$

In summary, this delivery disutility at the destination point of distribution includes:

- the initial production disutility (the factory gate disutility in its original production zone) of the good
- the cumulative disutility of all upstream stages of transporting the good via any intermediate DCs, through to the current DC
- the cumulative disutility of all associated operations in all the upstream DCs through which the goods have passed, prior to this current distribution stage

Multimodal Assignment

Motivation

The specific challenge for traffic assignment in a wide-area model is the presence of long-distance trips for both passengers and freight. Any long-distance trip may combine multiple modes from door to door. TRIMODE mode choice abstracts from this level of detail and adopts a main mode approach: each member of the choice set (Car, Coach, Rail, Air, Ship, ...) represents the choice of the dominant mode (or *main mode*) for the journey. Each main mode is associated with a set of *feeder modes*, which can be used in addition to the main mode.

Table 1. Selected Freight Main Modes with Feeder Modes in TRIMODE

Main mode	Feeder modes in hierarchical order
Air freight	1: Truck
Pipeline	1: Ship; 2: IWW ship; 3: Rail intermodal; 4: Truck
Inland waterway (IWW) ship	1: Rail; 2: Truck
Deep-sea ship	1: IWW ship; 2: Rail; 3: Truck
Rail	1: Truck
Truck	na

Table 1 lists some main modes in the TRIMODE model, with associated feeders.

Example: Deep-sea ship may be available as a main mode alternative, even though neither the origin nor destination zone has a deep-sea port, because a combined Truck—Deep-sea ship—Rail movement still counts as a trip of main mode Deep-sea ship. Each contiguous portion (u, v, m) of a trip using a single mode m from location u to location v is called a *mode leg*. The example trip comprises three mode legs (of modes Truck, Deep-sea ship, Rail). The *mode sequence* of a trip is the ordered set of its mode legs. In contrast to main modes, each mode (main or feeder) viewed in isolation is called a *single mode*.

Related Work

Combining multiple modes for a single trip is not novel. In the most common case, multiple public transport submodes are combined in one door-to-door journey. Two modeling approaches exist for this problem: either submode choice is treated as part of route choice in a supernetwork for all submodes (e.g., in Friedrich and Wekeck or Brands, de Romph, Veitch and Cook [*9*, *10*]) or mode choice is nested for the submodes (e.g., the two case studies in Gentile and Noekel [*11*]). These approaches fail when modes from public and private transport are combined, for example, in park-and-ride, because commonly available assignment procedures do not treat both simultaneously. The standard approach in this case involves matrix convolution on the costs skimmed from separate assignments, as described in Friedrich and Wekeck (*9*). Applications with longer sequences of modes arise in tour-based demand models where the concept of filter

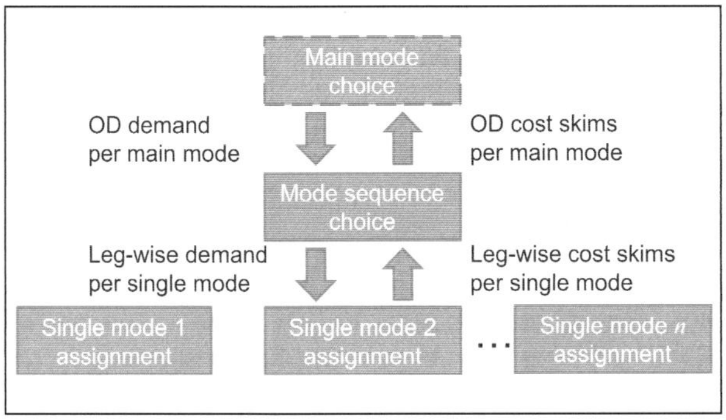

Figure 2. Bi-level formulation of TRIMODE assignment.

constraints on a reasonable order of the modes was first developed (e.g., in Hicks et al. or Roorda et al. [*12*, *13*]).

Formulation as a Bi-Level Problem

The approach presented here solves a generalized form of the park-and-ride assignment problem. Demand for a main mode from an origin to a destination is assigned to path alternatives in which each path is composed of one or more modal legs, using at a minimum the main mode, possibly in combination with feeder modes.

The multimodal assignment is formulated as a bi-level problem (Figure 2):

- At the upper level (mode sequence choice) the choice is between mode sequences for a given main mode. O-D demand for the main mode is received from the main mode choice (part of the demand model). The choice set for each O-D pair consists of feasible and efficient mode sequences (defined below). Generalized costs for each alternative are computed based on single-mode specific costs skimmed from the single-mode assignments (lower level). Total demand for the O-D pair is distributed over these alternatives, using a discrete choice model. Finally, for each single mode, demand is summed over all O-D pairs and over all mode legs that use this mode. This leg-wise single-mode demand is passed to the lower level. The volume-weighted average generalized cost of the mode sequences is passed as a main mode generalized cost to the main mode choice.
- At the lower level the leg-wise demand is assigned separately for each single mode (*single mode assignments*). Link volumes and paths are stored for each assignment. Generalized cost components are skimmed off each single-mode assignment and passed to the upper level.

As some of the cost components fed back from the lower to the upper level are flow-dependent, the bi-level problem is solved iteratively until a termination condition is met. Costs are suitably smoothed before being fed back, in order to improve convergence.

Mode Sequence Choice

Mode sequence choice is carried out on a graph that uses the same zoning structure as for the lower level, but abstracts from any link network detail. Mode sequence choice involves five steps:

1. Mode sequence search
2. Mode sequence utility calculation
3. Demand split across alternative mode sequences
4. Extraction of leg-wise single-mode demand
5. Extraction of main mode generalized cost skims

A mode sequence search is carried out on a multigraph $G = (Z, E), E \subseteq Z \times Z \times M$. Let $e-$, $e+$, m(e) denote the first, second, and third components of edge e. Z denotes the set of all zones. Zone indices o, d refer to the origin and destination zone (NUTS 3) of a complete trip. Indices u, v refer to the start and end zone (NUTS 3) of a mode leg. M denotes the set of all single modes, $\hat{M}$ the set of main modes. The procedure described below is applied to each main mode and each user class, but for brevity their indices are omitted in the equations.

Two important simplifications are introduced:

- Transfer locations between mode legs are also represented by zones.
- Mode legs within a mode sequence must join up at the transfer zones, but are implicitly assumed to fit together in time. Expected transfer waiting times and costs (if any) are assumed to be included in the leg-wise costs.

With these simplifications, any edge in G corresponds to a mode leg, and any path corresponds to a mode sequence.

From the single-mode assignments, skims ${}_h c^{m(e)}_{e-,e+}$ are available for each mode m and cost type $h \in H$, where H is the set of all cost types (time, distance, monetary cost, etc.). They induce edge costs ${}_h c(e)$:

$$ {}_h c(e) = {}_h c^{m(e)}_{e-,e+} \tag{13} $$

For path $p = (e_1, \ldots, e_n)$ we define:

$$ {}_h c(p) = \sum_{i=1}^{n} {}_h c(e_i) \tag{14} $$

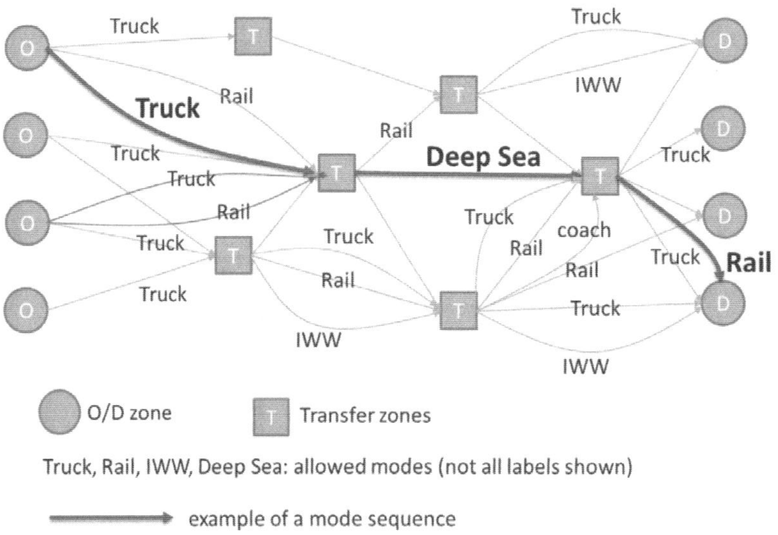

Figure 3. Example of a multigraph for mode sequence search, one mode sequence highlighted.

by summing mode specific cost over all mode legs. Loading, unloading, and storage costs and times at terminals are included in ${}_h c^{m(e)}_{e-,e+}$. Alternatively, zonal cost terms could be included in the summation of Equation 14.

From the individual cost types generalized costs are defined:

$$c(p) = \sum_h {}_h\beta_h c(p) \tag{15}$$

Only edges e with $c(e) < \infty$ belong to E. In other words, the edge from zone u to zone v using m exists if single mode m unimodally serves $u-v$.

Figure 3 displays an example of the multigraph and one mode sequence. The mode label is shown for some edges only. All nodes represent zones. In the example, the circles take the role of origins and destinations, boxes represent transfers.

Not all paths in G are realistic mode sequences. An unconstrained path search generates, for example, paths with many repetitions of Rail—Truck—Rail—Truck—Rail—Truck—... . Even if most of these unrealistic choice alternatives carried unattractive generalized costs and therefore attracted zero demand, they would greatly increase computational burden.

Constraints are therefore specified on the order in which feeder modes appear in a mode sequence. Table 1 lists the feeder modes in hierarchical order. Example: for main mode Deep-sea ship the highest-ranking feeder mode is IWW ship, the lowest-ranking feeder mode is Truck.

The following constraints apply:

1) The main mode must appear exactly once in a mode sequence.
2) Any feeder mode must appear at most once before the main mode and at most once after the main mode. This mode leg may contain transfers.
3) If a feeder mode appears *before* the main mode, it does so *after* all lower-ranking and *before* all higher-ranking feeders.
4) If a feeder mode appears *after* the main mode, it does so *before* all lower-ranking and *after* all higher-ranking feeders.

Example: mode sequence Rail—Air—Rail—Car satisfies the mode order constraint for main mode Air, whereas Rail—Car—Air does not. Rationale: it is considered unlikely that Rail is used to access a rental location for a Car that is then driven to the airport, so this sequence is ruled out.

A path $p = (e_1, \ldots, e_n)$ from o to d in G is *feasible* for main mode $\hat{m}$, if the mode sequence $(m(e_1), \ldots, m(e_n))$ of p satisfies the mode order constraint for $\hat{m}$.

A path $p = (e_1, \ldots, e_n)$ from o to d in G is *reasonable* for main mode $\hat{m}$, if Equation 16 and Equation 17 conditions apply:

$${}_{\hat{d}}c(p) \leq a \cdot {}_{\hat{d}}c(p^0) + b \tag{16}$$

where $a, b \geq 0$ are parameters and p^0 is a least-cost path from o to d using $\hat{m}$. Parameters a, b need to be calibrated to match empirically observed multimodal path choice sets:

$$\sum_{\substack{i=1 \\ m(e_i) \neq \hat{m}}}^{n} dd_{e_i-, e_i+} < \theta \cdot dd_{od} \tag{17}$$

where dd_{ij} is the straight-line (direct) distance between zones i and j.

Dominance criterion (Equation 16) is used to prune path alternatives that are too inferior to other alternatives. The chosen formulation has been used successfully in other assignment methods using branch-and-bound path search, including stochastic highway assignment and transit assignment (*14*, *15*). The Equation 17 condition eliminates path sequences in which the main mode is only used for an insignificant portion of the whole trip. Realistic results were obtained in numerical experiments with a threshold factor $\theta = 0.3..0.5$.

Complete enumeration of all mode sequences is impractical due to combinatorial explosion. Instead branch-and-bound search is run from each origin zone to all destination zones, keeping multiple paths as long as they are feasible and reasonable. The search is implemented using the approach described in *Base Year Freight Matrices: Final Report* (*4*) for connection search in schedule-based public transport assignment, except that here the time dimension is not considered.

Once the path choice set $P^{\hat{m}}_{od}$ from o to d for main mode $\hat{m}$ is constructed, total demand $OD^{\hat{m}}_{od}$ for the main mode is split across the alternatives. A logit model is used:

$$\pi^{\hat{m}}_{od}(p) = \frac{e^{\lambda c^{\hat{m}}_{od}(p)}}{\sum_{r \in P^{\hat{m}}_{od}} e^{\lambda c^{\hat{m}}_{od}(r)}} \tag{18}$$

$$OD^{\hat{m}}_{od}(p) = \pi^{\hat{m}}_{od}(p) \cdot OD^{\hat{m}}_{od} \tag{19}$$

Leg-wise demand is aggregated by single mode m by summing demand over all loaded mode sequences of all main modes that contain a leg of mode m. OD^{m}_{uv} is passed to the single-mode assignments at the lower level. Note that although $OD^{\hat{m}}_{od}(p)$ is given in natural units of the commodity transported (e.g., tons, containers, etc.), the input to the single-mode assignment typically needs to be in vehicles (e.g., trucks). The unit conversion of OD^{m}_{uv} using mode specific load factors is omitted here for brevity:

$$OD^{m}_{uv} = \sum_{\hat{m}} \sum_{\substack{p \in P^{\hat{m}}_{od} \\ (u,v,m) \in p}} OD^{\hat{m}}_{od}(p) \tag{20}$$

Finally, main mode costs ${}_h c^{\hat{m}}_{od}$ for all cost types h are computed as volume-weighted averages over all mode sequences for a given O-D pair. They are fed back to the main mode choice within the demand model:

$${}_h c^{\hat{m}}_{od} = \sum_{p \in P^{\hat{m}}_{od}} \pi^{\hat{m}}_{od}(p)\, {}_h c(p) \tag{21}$$

Single-Mode Assignments

After mode sequence choice, separate assignments for each single mode are run, all in parallel. The assignment method for each mode can be chosen independently, based on the characteristics of the mode, in particular drawing from both public and private transport assignment methods.

For all road-based single modes one multi-class equilibrium assignment is computed. For most discontinuous transport modes some form of public transport assignment is applied. The choice between schedule-based and frequency-based assignment depends on the frequency and regularity of the supply.

Where no explicit link network exists (e.g., bulk shipping services), only the resulting single-mode demand is stored in matrix form, without assignment to a network.

Intrazonal Movements Modeling

The spatial distribution task within the freight distribution channel choice model allocates the demand consumed in each destination NUTS 3 zone among the available origin zones for a particular logistic leg type. One potential origin is the destination zone itself and for some tertiary logistic legs it is expected that such intrazonal movements will be the majority of total movements. At the same time, for passenger demand a large share of trips is intrazonal. Given the size of NUTS 3 regions, the length of intrazonal movements can be of a few kilometers as well as of 20 km or more. In order to represent intrazonal transport disutility more accurately, and also to model availability and competitiveness of alternative vehicle types, intrazonal freight movements are further distributed among four different distance bands, namely:

- Urban movements short
- Urban movements long
- Rural movements short
- Rural movements long

This distance band approach to intrazonal freight modeling is an adaptation of that originally developed for use in the SCENES model (*16*) of Europe to ensure that it represented all freight and passenger travel demand on all modes, despite SCENES being implemented with a NUTS 2 level of zoning (NUTS 2 regions are aggregations of NUTS 3 regions; there are 276 NUTS 2 regions in the EU). This methodology has since been refined by combining it with an area-type-based zoning system to provide the core functionality underpinning the policy model component of the national passenger transport model (NTMUK) of Great Britain, which has been used intensively for policy assessment by the UK Department for Transport from 2002 through to the present.

The intrazonal distribution among distance bands within a zone is based on exogenous shares estimated separately by freight type, logistic leg, and zone type (in metropolitan areas intrazonal movements tend to be urban ones, whereas in sparsely populated areas rural movements prevail).

Within each distance band, freight activity is split among alternative road vehicle types using a logit algorithm:

$$ {}_{v}p_{id}^{rl} = \frac{e^{-\psi^{rl}{}_{v}u_{id}^{rl}}}{\sum_{v} e^{-\psi^{rl}{}_{v}u_{id}^{rl}}} \tag{22}$$

where

${}_{v}u_{id}^{rl}$ is the disutility of delivering a unit of freight along distance band d within an intrazonal movement in zone i for freight type r and logistic leg l.

ψ^{rl} are calibrated parameters that denote the extent to which intrazonal trips or deliveries of freight type r and logistic leg l concentrate within those vehicle types with the lowest total disutility of delivery.

The vehicle transport costs and speeds that are used to compute disutility in each distance band are exogenous inputs, again differentiated by the area type of the zone.

For the purposes of distribution, the average zone disutility is estimated as a weighted average of the logsum of vehicle types' disutilities:

$$ u_{i}^{rl} = \sum_{d}\left[-\frac{1}{\psi^{rl}}*\ln\left(\sum_{v} e^{-\psi^{rl}{}_{v}u_{id}^{rl}}\right) * DS_{zd}^{rl}\right] \tag{23}$$

where the term DS_{zd}^{rl} represents the exogenous shares mentioned above, measuring how the intrazonal freight demand of each specific segment is distributed among the four distance bands.

The reason for developing this distance band methodology is to enable the supply characteristics of short-distance freight movements to be represented adequately, while avoiding excessive computing costs. Model run times generally increase in proportion to the square of the total number of zones. However, this distance band approach alleviates the need to introduce a much larger number of small zones and so generates minimal extra computing burden, while providing a complete representation of total travel.

Conclusions

The methodological developments introduced above have been designed to improve the realism and performance of comprehensive operational freight modeling procedures for large study areas.

The impact of trends in logistics and of the responsiveness of logistics to policy initiatives can best be understood realistically, if these impacts are considered within the context of both their underlying P-C and O-D matrices. The cost-effectiveness of individual distribution channels is highly sensitive to transport and logistics cost structures so the resulting spatial pattern of O-D movements should take major account of this sensitivity. The distribution channel choice procedure provides a computationally efficient method for representing the complexity of the operation of logistics and for representing how the distribution structure varies greatly across broad types of goods and for representing how it will evolve in response to cost changes.

The greatest challenge in the implementation of the distribution channel choice lies in obtaining access to comprehensive data that describes the actual scale of usage of competing distribution channels. A major data assembly and consistency analysis task, combining Eurostat European data with individual country data, is underway to provide a solid foundation for estimating the diverse set of parameter values required within TRIMODE. The residual disutility-based methods for constraint matching have been developed in part to ensure that flexible and effective use can be made of whatever aggregate data does become available.

The ability to provide a consistent representation of the door-to-door costs for the choice set of multimodal options available for each O-D pair is important in ensuring that appropriate transport costs are input into the distribution choice procedure. To achieve this, the bi-level formulation for multimodal assignment presented here has a number of advantages in practical application:

- Mode sequence choice for all main modes can run in parallel, as can all single-mode assignments, reducing execution time and allowing distributed computation.
- Once leg-wise single-mode demand has been calculated for all road-based single modes (e.g., Car and Truck), it is passed into a single multi-class highway assignment at the lower level. Equilibration between Car and Truck is achieved without further outer loops.

- The simplifying assumption that transfer locations are represented by zones reduces network coding effort, as there is no need to create transfer links between single-mode networks.

Some limitations should, however, be noted for the formulation presented here:

- Large zones may connect to different network nodes for different modes, in which case inter-mode transfer costs at such a zone may be incorrectly represented.
- In a wide-area model like TRIMODE it may be justified to neglect the question whether mode legs join up in time. The same would not be true in urban applications, for example, for the combination of shared mobility services with conventional public transport. Adding the time dimension would increase the computational burden, however, and needs to be investigated further.

The use of distance bands for intrazonal movements adds flexibility and ensures that the average travel length of local distribution consignments can depend on zonal features like density. It enables the characteristics and responsiveness of shorter freight trips to be represented in a realistic fashion without requiring the adoption of a much larger number of small zones that would in turn generate infeasible computing run times.

Validation and calibration of the model against data from European and national freight statistics and surveys is part of the TRIMODE project. Work on this is starting at the time of writing. The model is expected to be completed by mid-2019.

Author Contributions

The authors confirm contribution to the paper as follows: freight distribution channel modeling: IW; multimodal assignment: KN; intrazonal movements modeling: DF; overall model architecture: KN, IW, DF; model implementation: KN, DF. All authors reviewed the results and approved the final version of the manuscript.

References

1. EUROSTAT. *NUTS: Nomenclature of Territorial Units for Statistics*. http://ec.europa.eu/eurostat/web/nuts.
2. European Commission. *GEM-E3 General Equilibrium Model for Economy: Energy—Environment*. https://ec.europa.eu/jrc/en/gem-e3/model. Accessed July 15, 2017.
3. Williams, I. N., and M. H. Echenique. A Regional Model for Commodity and Passenger Flows. *Proc., PTRC Summer Annual Meeting, Stream F*, PTRC Education and Research Services Ltd, London, 1978, pp. 121–128.
4. Jin, Y., I. Williams, and M. Shahkarami. Integrated Regional Economic and Freight Logistics Modelling: Results from a Model for the Trans-Pennine Corridor, UK. Paper to the European Transport Conference, 2005. http://abstracts.aetransport.org/paper/index/id/2207/confid/11. Accessed June 3, 2017.
5. WSP. *Base Year Freight Matrices: Final Report*. For Department for Transport, Contract PPRO 04/03/14, 2011. http://data.dft.gov.uk/byfm/byfm-report.pdf. Accessed June 3, 2017.
6. Huber, S., J. Klauenberg, and C. Thaller. Consideration of Transport Logistics Hubs in Freight Transport Demand Models. *European Transport Research Review*, Vol. 7, No. 4, 2015, p. 32.
7. de Jong, G., M. Ben-Akiva, J. Baak, and S. E. Grønland. Method Report: Logistics Model in the Norwegian National Freight Model System (Version 3), 2013. http://www.ntp.dep.no/Transportanalyser/Transportanalyse + godstransport/. Accessed June 3, 2017.
8. Kawamura, K., A. K. Mohammadian, A. Samimi, Z. Pourabdollahi, and M. Klekotka. *Development of a Freight Policy Analysis Tool for Northeastern Illinois and the United States: Freight Activity Microsimulation Estimator*. Urban Transportation Center, The University of Illinois at Chicago, Chicago, IL, 2011. http://www.idot.illinois.gov/assets/uploads/files/transportation-system/reports/dpit/final%20report%20freight%20activity%20microsimulation%20estimator%20(fame).pdf. Accessed June 3, 2017.
9. Friedrich, M., and S. Wekeck. A Schedule-Based Transit Assignment Model Addressing the Passengers' Choice Among Competing Connections. In *Schedule-Based Dynamic Transit Modeling: Theory and Applications*, Kluwer Academic Publishers, Dordrecht, The Netherlands, 2004, pp. 159–173.
10. Brands, T., E. de Romph, T. Veitch, and J. Cook. Modelling Public Transport Route Choice, with Multiple Access and Egress modes. *Transportation Research Procedia*, Vol. 1, 2014, pp. 12–23.
11. Gentile, G., and K. Noekel, eds. *Modelling Public Transport Passenger Flows in the Era of Intelligent Transport Systems*. Springer, Heidelberg, Germany, 2016.
12. Vovsha, P., J. Hicks, G. Vyas, V. Livshits, K. Jeon, R. Anderson, and G. Giaimo. Combinatorial Tour Mode Choice. Presented at 96th Annual Meeting of the Transportation Research Board, Washington, D.C., 2017.
13. Roorda, M. J., D. Passmore, and E. J. Miller. Including Minor Modes of Transport in a Tour-Based Mode Choice Model with Household Interactions. *Journal of Transportation Engineering*, Vol. 135, No. 12, 2009, pp. 935–945.
14. Spiess, H. *Computing Activity Chain Based Trip Distribution Models*. EMME/2 Support Center, CH-2558, Aegerten, Switzerland 1993.
15. PTV Group. *PTV Visum 17 Manual*. PTV Group, Karlsruhe, Germany, 2017.
16. ME&P, TRT. *SCENES European Transport Forecasting and Appended Module: Technical Description, Deliverable D4 of the SCENES*. Project Funded by the European Commission Under the Transport RTD Programme of the 4th Framework Programme, 2000.

The Standing Committee on Freight Transportation Data (ABJ90) peer-reviewed this paper (18-02449).